TURN-OF-THE-CENTURY
FARM TOOLS
AND
IMPLEMENTS
PETER HENDERSON & CO.

With a New Introduction by
Victor M. Linoff

DOVER PUBLICATIONS, INC.
MINEOLA, NEW YORK

Copyright

Copyright © 2002 by Dover Publications, Inc.
All rights reserved under Pan American and International Copyright Conventions.

Published in the United Kingdom by David & Charles, Brunel House, Forde Close, Newton Abbot, Devon TQ12 4PU.

Bibliographical Note

This Dover edition, first published in 2002, is an unabridged republication of the *Catalogue of Tools and Implements, Fertilizers, Insecticides and Essentials . . .* originally published by Peter Henderson & Co., New York, 1898. A new introduction has been specially prepared for this edition.

DOVER *Pictorial Archive* SERIES

Library of Congress Cataloging-in-Publication Data

Catalogue of tools and implements, fertilizers, insecticides, and essentials.
    Turn of the century farm tools and implements / Peter Henderson & Co.
        p. cm. — (Dover pictorial archive series)
    Originally published: Catalogue of tools and implements, fertilizers, insecticides, and essentials. New York : P. Henderson, 1898. With a new introd.
    ISBN 0-486-42114-7 (pbk.)
       1. Agricultural implements—United States—Catalogs. 2. Farm equipment—United States—Catalogs. I. Peter Henderson & Co. II. Title. III. Series.
S676.3 .C38 2002
631.3—dc21

2002024730

Manufactured in the United States of America
Dover Publications, Inc., 31 East 2nd Street, Mineola, N.Y. 11501

INTRODUCTION

By Victor M. Linoff

The Garden

While the origins of gardening are shrouded in history, growing flowers, fruits, and vegetables is an ancient activity, one that no doubt dates back to the time when man turned from a nomadic hunting and gathering way of life to staying in one place and cultivating his own food. As life became more settled, the ability to produce a steady supply of food by sowing crops enabled humans to pursue less strictly utilitarian endeavors. So it was that gardening came to include both the practical and the ornamental.

As the centuries passed, gardening continued to grow in popularity, resulting in the development of a literature of gardening. In fact, writings on the subject date back at least as far as the sixteenth century.[1] By the nineteenth century, gardening had become an integral part of Victorian life. Books and periodicals proliferated, offering the latest information, trends, and advice about all manner of gardens and crafts.

When the Henderson Company Catalogue was published in 1898, gardening had evolved into a multifaceted activity: social, recreational, and commercial. For many, market and/or truck gardening was either a livelihood or an additional source of income.[2] Peter Henderson has been credited with being an important influence on this kind of commercial gardening. After the Civil War, many returning veterans, reading his newly published *Gardening for Profit,* were encouraged to try market gardening.[3] Others merely gardened for pleasure. Recreational and ornamental gardening was no longer the exclusive prerogative of the wealthy; people of all classes and in all situations were actively gardening for pleasure and fun.

In addition to sheer beauty, gardens provided fruits and vegetables, herbal medicines and spices, and an abundant supply of fresh cut and aromatic flowers for decorating the home. In urban settings where outdoor gardens weren't generally possible, window planter boxes were frequently used. Larger homes often featured greenhouses and conservatories. The nineteenth century was an era of great ingenuity and diverse attempts to bring the glory of nature indoors. This penchant for the natural was manifested in many ways. Drying flowers for crafts and art projects was a popular pastime. Floral prints, nature studies, and still lifes of fruit and flowers abounded. Floral and plant motifs were ubiquitous ornamentation on furniture, wallpaper, and other surfaces. In fact, the Art Nouveau style, with its sinuous and luxuriant forms, was the ultimate homage Victorians paid to the beauty of nature.

Gardening on a large scale, i.e. farming, was one of the largest nineteenth-century industries. "There were, at the end of 1868, in the United States 2,033,665 farms, comprising 405,280,851 acres—an average of 199 acres for each farm."[4] Add to that countless private and commercial urban gardens and it's easy to understand why more than 600 U.S. companies offered seeds, tools, implements, and supplies for recreational, market, and truck gardening and farming. Along with the Peter Henderson Company, major firms whose names are still recognized today include the W. Atlee Burpee Company, Ferry & Co., Chas. C. Hart Seed Company, James John Howard Gregory, and James Vick. Large general catalog houses like Sears, Roebuck & Co. and Montgomery Ward were also major providers of materials.

The Man and His Company

During his lifetime Peter Henderson established himself as a leader in the field of horticulture and floriculture. Though never formally trained, the mostly self-taught Henderson became one of the most recognized and respected gardening authorities and businessmen in the U.S. So it is rather surprising that little has been recorded about Peter Henderson the man. Most of what is known comes from an emotional memoir and tribute to his father written by his son Alfred, also a partner in the business, upon Henderson's death in 1890.[5]

The youngest of three children, Peter Henderson was born in 1822 in Pathead, Scotland, a small farming village twelve miles from Edinburgh. His father James, a local land steward, essentially raised him after his mother Agnes died when Peter was eight. Moreover, though he died twelve years before Henderson's birth, it is speculated that his maternal grandfather (Peter Gilchrist, 1740–1810) was a major influence on the young man's choice of career. Gilchrist had been a respected nurseryman and florist, whose collection of horticultural writings and other materials undoubtedly came to the attention of the budding gardener.

From age sixteen to twenty, Henderson was apprenticed to George Sterling, head gardener at Melville Castle near Dalkeith. Young Peter showed a real interest in, and proclivity for his work. Sterling saw a special gift in his apprentice and as a mentor became a powerful influence in his life. After just one year under Sterling's tutelage, Henderson competed for and won the first of many awards, a medal from the Royal Botanical Society of Edinburgh for the "…best herbarium of native and exotic plants."[6]

At age 21, apparently following in the path of his brother, James, Peter Henderson emigrated to the United States in 1843. He immediately began working in nurseries; first in Astoria, Long Island, and then in Philadelphia where he was employed by Robert Buist, Sr., a leading nurseryman of the day. That experience led to employment as a private gardener in Pittsburgh.

By 1847, the resourceful 25-year-old had saved $500—enough to form a partnership with his brother James in Jersey City, New Jersey. Together they worked on a rented ten-acre plot with three small greenhouses. Several years later James went off on his own to concentrate on vegetable gardening. Peter Henderson continued at the Jersey City site until 1864. Over time he had acquired about ten acres a mile away in South Bergen. On this land Henderson "…erected what was at the time considered a model range of greenhouses, heated and ventilated in the best known methods then in vogue."[7] Although there would be a number of changes and partners over the years, this venture marked the start of a business that would endure for more than a century. (See chronology on page vi)

On the personal front, Henderson met and married Emily Gibbons, a native of England, in 1851. Together they raised two sons and a daughter. Three years after Emily's untimely death in 1868 at age 34, Henderson remarried. His second wife, Jean Reid, was the daughter of his friend and former partner, Andrew Reid. She survived Henderson's passing in 1890.

A decade after his arrival in the U.S., Peter Henderson began his first retail activity in New York City. He shared the offices of McIlvain & Orr, produce merchants and auctioneers. During the spring and summer growing season, Henderson would take daily orders for fresh fruits and vegetables that would be delivered the next day from his greenhouses and gardens just across the Hudson River in Jersey City. Additionally, McIlvain & Orr would auction Henderson's surplus produce and seeds.

Their relationship lasted until 1862, when Henderson joined two young Scottish gardeners who had just opened a seed store at 67 Nassau Street. It was at that time that Henderson began advertising and producing catalogs. In 1865, he purchased the interests of one of the partners. The company then became known as Henderson & Fleming.

That partnership was dissolved in 1871. Henderson once again relocated his retail produce business and began selling seeds commercially at 35 & 37 Cortlandt Street in New York City. This address would be home to the newly named Peter Henderson & Company for more than 75 years.

Even while actively pursuing his business interests, Henderson was always an avid researcher and experimenter, frequently contributing to the popular periodicals of the day.[8] *American Agriculturist, Gardener's Monthly, The Horticulturist, Rural New Yorker,* and *Country Gentleman* all featured articles by Henderson. Those essays eventually led to his first book, *Gardening for Profit,* published in 1867. Penned by a reluctant Henderson during the summer of 1866 at the urging of several colleagues, the book was written mostly during breaks in Henderson's typical sixteen-hour workdays. He completed the book in a remarkable 100 hours. At the time of his death, *Gardening for Profit* was in its forty-first edition and had been revised four times to accommodate changing technology. Today it is still considered a seminal work.

Five more books, numerous monographs and articles followed.[9] Indeed, Henderson continued writing up until a month before his death. His writings were widely available through his catalogs. The 1898 catalogue, for example, makes an inviting offer to the reader: ten Henderson books and monographs for ten dollars.

A man who enjoyed challenging conventional thinking, Peter Henderson relished a good debate. In fact, he even took on Charles Darwin over a couple of issues. The most celebrated disagreement dealt with a plant called the Carolina flytrap. Darwin contended that nourishment of the plant came from the insects it trapped. Henderson conducted a series of experiments designed to disprove Darwin's assertion, claiming that the plant's sustenance came from water and soil. Only years later, after Henderson's death, did the Darwinian conjecture prevail.[10]

Like other contemporary leading businessmen, Henderson saw his future growth inextricably tied to national distribution. With the advent of rail, nationwide delivery of goods became fast and easy. Henderson also understood the value of catalogs and advertising as a powerful tool for expanding his business. By 1872, he had developed a distinctive marketing plan. He created the tagline "Everything for the Garden," promoting the idea that all the necessary supplies for successful gardening could be acquired from him at one location.[11] A writer with an easy-to-read conversational style, Henderson wrote all the editorial copy and descriptions for his catalogs until 1880.

It was said that this kindly man who retained his quaint Scottish brogue took a genuine interest in people. He mixed and made friends as easily with renowned figures like Andrew Carnegie and the Reverend Henry Ward Beecher as with rank and file workers and the general public. For most of his career he personally answered, often in longhand, the large volume of correspondence directed to him.

On January 17, 1890, after a short bout with pneumonia, the energetic Peter Henderson died quietly in his Jersey City home of more than forty years. His bed overlooked the beloved gardens and hothouses that had for so long been integral to his life. Testifying to the regard in which he was held, his family "...received from all parts of the world, nearly 8000 letters of sympathy and condolence" in the eleven months following Henderson's death.[12]

Peter Henderson & Company had survived more than a century of changing technology, economics and business practices, but by the end of World War II and the consequent waning of the Victory Garden, which had revived interest in home gardens, the venerable old company was in the throes of an irreversible decline.[13] In an effort to survive, Henderson & Co. merged in 1951 with Stumpp & Walter, a longtime seed marketer and equipment supplier. Like its new partner, Stumpp & Walter was struggling to stay alive in changing times. Two years later bankruptcy brought down their alliance and ended forever the 106-year run of Henderson & Company. Yet the memory of its founder, and his pioneering contributions to horticulture still remain in the prolific writings of Peter Henderson, and in his catalogs.

The Catalogue

While this catalogue, issued in 1898, was published eight years after Henderson's death, it nonetheless bears his distinctive imprint in terms of philosophy, design, and layout. Within its 72 pages can be found nearly everything for successful growing and cultivation. As the cover copy suggests, the catalogue attempted to supply every need of the market it served, and it's precisely this diversity that makes the catalogue so valuable to students of Victorian horticulture. In effect, it offers a comprehensive inventory of late Victorian tools, accessories, and decorative items for every manner of gardening, farming, or animal husbandry. Moreover, nearly every item is illustrated, from the horse-driven Farmer's Corn and Cob Mill to Mann's Superior Bone Cutters.[14]

Much of the catalogue is directed toward farmers and commercial growing interests. Plows, harrows, manure spreaders, planters, and seeders take up the first fifteen pages, while the needs of the small gardener or grower occupy the central portion of the publication.

Considerable attention is paid to lawns, which had become an important feature in Victorian landscaping.[15] Mowers of all sorts and ornamental and practical sprinklers abound. Henderson even addressed the international market by offering The English Golf Mower "made especially for British Golf Links." (p. 47).

In addition to the standard implements of the day, the catalogue contains numerous examples of fanciful Victorian ingenuity. For $17.50, a farmer could utilize a dog, sheep, goat, or calf to provide the motive force for a Railway or Tread Power (p. 31). And, of course, every good gardener could improve his efficiency with a Watts Asparagus Buncher or a Box Pattern Horse Radish Grater (p. 34). The Aspinwall Potato Cutter made the perfect accessory to the Aspinwall Potato Planter (p. 18). The combination of the Family Pop Corn Sheller and the "Acme" Corn Popper provided households with an easy way to enjoy this popular treat (p. 43).

Victorian gadgets abound, including ingenious devices for dealing with such garden pests as moles. A highly effective way of dealing with moles is graphically portrayed on page 55.

Pages 42 and 54 depict delightful Victorian garden furniture, including some wonderful twisted wire plant stands and vine trainers that were very much the rage during the last half of the nineteenth century.

The catalogue provided extra value with its extensive bibliography of useful "books on horticulture, agriculture and kindred subjects" appearing on pages 2–4.

Some of the finest catalogs ever produced came from seed and garden suppliers. In addition to implements and tools, Peter Henderson & Co. also sold seeds. It should be noted that while seed catalogs were more often saved and are still fairly abundant today, implement catalogues illustrating the varied and sometimes arcane equipment used by Victorian-era horticulturists and agriculturists are considerably more scarce. Hence the significance of this 1898 record of gardening history.

Henderson Associations and Chronology

While additional research is necessary to fully delineate the exact chronology of the 106 years of the various Henderson Company operations, what follows can serve as a basic outline of Peter Henderson's business activity.

1847-185? Partnership with brother James, Jersey City, NJ

McIlvain & Orr, No. 9 John Street, New York City

1864? Operated seed business, greenhouses, and garden on 10 acres on Arlington Avenue,
    Bergen District, Jersey City, NJ

(James) Fleming & (William J.) Davidson, 67 Nassau Street, New York City

Henderson & Fleming, 67 Nassau Street, New York City

Peter Henderson & Co. (Partners: Peter Henderson, William H. Carson, and Alfred Henderson),
    35 & 37 Cortlandt Street, New York City

Peter Henderson & Co. (Partners: Peter Henderson, James Reid, and Alfred Henderson),
    35 & 37 Cortlandt Street, New York City

Peter Henderson & Co. (Partners: Peter Henderson, Alfred and Charles Henderson),
    35 & 37 Cortlandt Street, New York City

Peter Henderson & Co., 35 & 37 Cortlandt Street, New York City

1951-1953 Henderson, Stumpp & Walter, New York City

Endnotes

1   Published in 1577, *The Gardener's Labyrinth* by Thomas Hyll offered strategies for successful growing. R. Gardiner wrote a treatise called *Instructions for Kitchen Gardens* in 1603.

2   In Victorian times a *farm* was considered a rural plot of land "cultivated for the purpose of profit." Gardens were generally a smaller "piece of ground enclosed and appropriated to the cultivation of herbs & plants, fruits & flowers." [L. Colange, (see footnote 4 below)] The terms *market* and *truck* gardening were Victorian descriptions of distinct growing and selling strategies. *Market Gardening* refers to the proximity of a produce or flower garden to the point of sale. In the era before refrigeration and cold storage, perishable produce, by necessity, was often grown in small urban gardens where the output could be readily transported and sold.

   Truck Gardening or *Farming* implies that because greater distances exist between garden or field and the place of sale, produce must be transported. Robert F. Becker, Associate Professor of Horticultural Sciences at Cornell University suggests that the term *truck gardening* may have a two-fold meaning. The obvious one refers to the means of transporting produce to market. The other, more subtle, interpretation is that it comes from *troquer,* a French term for barter or exchange, a form of trade often used by farmers.

3   Joseph Kastner, Seed Centennial, *Life* magazine April 7, 1947. Throughout his writings Henderson discusses the commercial aspects of gardening. In *Gardening for Pleasure* he recognizes "that hundreds every season, who have a taste for horticulture, branch out from private into commercial gardening, either from necessity or for a love of making a business of the work." (p. 7)

4   L. Colange, *Zell's Universal Dictionary of English, Language, Science, Literature & Art,* T. Elwood Zell, Philadelphia, 1871

5   Alfred Henderson, *Peter Henderson, Gardener—Author—Merchant: A Memoir,* New York, 1890. This scarce 48-page memoir is reprinted in a Henderson anthology published by The American Botanist, Booksellers. See bibliography.

6 ibid (p. 13). A herbarium is a collection of dried plant specimens, mounted and arranged in an organized manner, as a reference source.

7 Alfred Henderson, op.cit. (p. 20)

8 While prior to World War II more than 75% of Henderson seeds were imported from Europe, Peter Henderson was credited in *Life* magazine (April 7, 1947) with introducing a wide variety of species including "the double zinnia (1865), the Early Snowball cauliflower (1878), the Premier pansy, Mignonette lettuce, the Ponderosa tomato, and the bush lima bean (1889)."

9 At the time of Henderson's death his six books had sold more than 250,000 copies.

10 *Life* magazine, *Gardening for Profit* reprint

11 ibid (p. 22)

12 Alfred Henderson, op.cit. (p. 8)

13 Nonetheless, its 100th anniversary was still an occasion for a *Life* magazine three-page paean to Henderson & Co.'s longevity.

14 It's interesting to observe that while the copy says "they may be operated by boy or man" the illustration depicts a well-dressed woman handily working the device.

15 Prior to the Civil War, few urban or rural American homes had lawns. Peter Henderson may have been one of the first to advocate well-kept lawns as a part of residential landscaping. The development of the lawn mower made it possible. In the U.S., the first patents for lawn mowers were granted in 1868. Over the next five years mower patents rose to thirty-eight. So, it is not surprising that in 1875 Henderson observed in *Gardening for Pleasure* that "Since the introduction of the lawn mower, the keeping of the lawn has been so simplified that no suburban residence is complete without one."

Bibliography and Selected Reading

Berrall, Julia S., *The Garden, An Illustrated History,* The Viking Press,
New York, 1966

Ewan, Joseph, ed., *A Short History of Botany in the United States,* Hafner Publishing Co.,
New York and London, 1969

Henderson, Alfred, *Peter Henderson, Gardener—Author—Merchant: A Memoir,*
Press of McIlroy & Emmet, New York, 1890.

Henderson, Peter,

Gardening for Profit: A Guide to the Successful Cultivation of the Market & Family Garden,
Orange, Judd & Co., New York, NY, 1867

Gardening for Pleasure, Orange, Judd & Co., New York, NY, 1887

Gardening for Profit. (Reprint of 1867 first edition. Also includes a history of market gardening in
America, selections from Henderson's *Garden & Farm Topics* and a reprint of *Peter Henderson,
Gardener—Author—Merchant: A Memoir* by Alfred Henderson, New York, 1890), The American
Botanist, Booksellers, Chillicothe, IL, 1991, 1997

Practical Floriculture: A Guide to Successful Cultivation of Florist's Plants,
Orange, Judd & Co., New York, NY, 1869

Iverson, Richard R., *The Exotic Garden,* Taunton Press, Newtown, CT, 1999

Jenkins, Virginia Scott, *The Lawn, A History of an American Obsession,* Smithsonian Institution,
Washington, DC, 1994

Kelley, Etna M., *The Business Founding Date Directory,* Scarsdale, NY, 1954

Leopold, Allison Kyle, *The Victorian Garden,* Clarkson Potter, New York, 1995

Periodicals

Life magazine, April 7, 1949
Scribner's Monthly, Vol. XXII, Issue 2, June, 1880, pp. 219–229

Acknowledgement

I am deeply indebted to Wendy Inman for her invaluable assistance in researching the history of the
Peter Henderson & Company and for her generosity in taking the time to review the manuscript for
accuracy.

Peter Henderson & Co.

35 & 37 Cortlandt Street, — NEW YORK.

Catalogue

..... OF .....

Tools and Implements,

FERTILIZERS, INSECTICIDES AND ESSENTIALS,

—) FOR THE (—

GARDEN, FARM, GREENHOUSE, LAWN, ORCHARD, POULTRY-YARD, STABLE & HOUSEHOLD.

(Original Cover)

INDEX

The prices quoted in this Catalogue are for the articles packed and delivered to transportation companies in New York, freight or expressage to be paid by consignee.

A
	PAGE
Anvil and Vise	34
Apple Parer	43
" Corers	43
Asparagus Buncher	34
" Knives	38
Axes	38
" Turf Axe	53
Axe Hatchet	39

B
Bag Holder	29
Balling Iron	61
Baling Presses	25, 31
Barrel Headers	30
Baskets	38
Bean Harvester	16
" Planter	16, 17
Bellows—Powder	66
Bill Hooks	38
Binder	24
" Twine	24
Bone Cutters	27
" Mill	29
Books	1 to 4
Border Trimmers	49, 53
Brackets for Pots	42
Brooms	38, 53
Broadcast Seeders	17
Brooders	59
Brush Axes	38
Budding Knives	39
Bull Leader	38
" Punch	38
Burning Iron	38
Butter Worker	44
" Printer	44
" Moulds	44
" Boxes	44

C
Calf Weaner	44
" Feeder	44
Caponizing Tools	57
Cards	38
Cattle Comfort	61
Carts for Horse Power	32
" " Hand "	33, 39
" Sprinkling	32, 52
Carriage Jack	34
Celery Hillers	13
Chain Ties, etc	38
Cherry Stoners	43
Chicken Coop	57
Churns	44
Clippers—Dehorning	44
" —Horse	40
Cider Mills	30
Cold Frames	37
Cookers	31, 43
Corn Planters	16, 17, 18
" Husker	38
" Knives	38
" Harvester	16
" Stalk Cutter	16
" Mills	29
" Popper	43
" Shellers	28
" Stalk Shredder	26
Coverer and Hiller	13
Cradles—Grain	38
Crowbars	38
Cultivators—	
Hand Power	19 to 23
Horse Power	10 to 12
Cultivator Harrows	8, 9, 11
Curry Combs	38
Creamers	45
Cream Gauges	45
" Separators	44, 45

D
	PAGE
Dibbers	38
Drain Cleaners	38
Drain Tile Layer	38
Drinking Fountains—	
Poultry	57

E
Egg Tester	57
" Carriers	57
Evaporators—Fruit	30

F
Fan Mills	29
Farmers' Kit of Blacksmith Tools	34
Feather Pulling Bit	57
Feed Cutters	26, 27
" Cooker	31
" Trays—Poultry	57
Fencing Wire, etc	36
Fertilizers	69
" Drills	15, 20
" Broadcaster	15, 49
" Distributers	56
Floral Tools	38
Fodder Shredder	26
Forge	34
Forks	38
Fruit Pickers	34, 38
" Presses	30, 43
" Evaporators	30
" Sorter	30
Fumigators	67
Fungicides	62

G
Gape Worm Extractor	57
Garden Line	38
" Reels	38
Garden Engine	34
Glazing Points	37
Glaziers' Diamond	37
Gloves	38
Grader	23, 49
Grape Seeder	43
Grafting Chisel	38
" Wax	38
Grass Hooks	53
Graters	34, 43
Grain Drills	17
Grindstones	34, 43
Grit Crusher	57
Grub Hoe	39
" Plow	7

H
Hanging Baskets	54, 55
" Pots	42
Halter Chains	38
Hammers	39
Harrows	8
Harness	32
" Oil	41
Hatchets	39
Hay Tedder	25
" Caps	25
" Rakes—Horse	25
" " —Hand	25, 40
" Knives	38
" Loader	25
Heaters—Oil	42
Hedge Shears	53
" Knife	39
Hens' Nests	57
Hoes—Hand	39
" —Wheel	19 to 23
Hog Ringer and Rings	39
" Scraper	39
Horse Hoes	10 to 12
" Grape Hoe	11
" Clippers	40

H
	PAGE
Horse Boots	47
Hose Rubber	52
" Nozzles	51
" Holder	51
" Reels	52
" Menders	51
" Couplers	51
Hot Bed Frames	37
" Sash	37
" Mats	37

I
Ice Cream Freezers	43
" Incubators	58
Insecticides	62

J
Jacks—Wagon, etc	34

K
Knapsack Sprayers	64
Knives—Asparagus	38
Knives—Corn	38
" —Hay	38
" —Hedge	39
" —Budding	39
" —Pruning	39
" —Poultry Killing	57

L
Labels—Plant and Tree	39
" —Poultry Marking	57
Lactometer	45
Lactoscope	45
Ladders	34
Lanterns	39
Lawn Lamps	39
" Carts	49
" Edgers	49, 53
" Enricher	55
" Feeder	56
" Grader	49
" Mowers	46, 47
" " Sharpener	48
" Rakes	53
" Rollers	14, 48
" Settees	54
" Sprinklers	50, 51
" Sweepers	48
" Shears	53
" Umbrella	54
Leveler—Road	23
" and Pulverizer	9
Lice Killing Machine	57

M
Manure Spreaders	15
" Hooks	40
Mangers and Racks	39
Marker	13
" for Poultry	57
Mastica	37
Mats—Hot Bed	37
Measures	39
Meat Choppers	43
Medicines for Stock	60, 61
Milk Coolers	45
" Tester	45
" Pails	45
" Strainers	45
" Separators	45
Milking Tubes	44
Mills—Bone	27, 29
" —Cider	30
" —Fan	29
" —Grinding	29
Mole Traps	55
Moisture Gauge	57
Mowers—Field	24
" —Lawn	46, 47
Mortar	57
Muzzles for Horses, etc	40

N
	PAGE
Nest Eggs	57

O
Oil Cans	40
Onion Drill	23
" Puller	23

P
Pails—Milk	45
" —Stable, etc	40
Pencils	40
Perch Brackets	57
Picks	40
Plant Setting Machine	18
" Protectors	37
" Sprinklers—Rubber	40
" Stands	42
Plows—Horse Power	5 to 7
" —Man Power	19 to 22
Post Hole Digger	31
" " Auger	31
" " Rammer	40
" " Spoon	41
Potato Bug Destroyers	32, 66
" Coverer	13
" Cutter	18
" Diggers	7, 18
" Hooks	40
" Planters	18
" Parers	43
" Slicers	43
" Sorter	18
Pots	42
Poultry Supplies	57 to 60
" Foods	60
" Medicines	60
Powers	31
Powder Distributers	66, 67
Pruners—Tree	40
" —Knives	39
" —Shears	40
Protecting Cloth	37
Pumps	34
" Spraying	64, 65
Putty Bulbs—Rubber	40

R
Rakes—Horse	25
" —Lawn	53
" —Garden	40
Raphia	41
Raisin Seeder	43
Reaper	24
Reels for Hose	52
" Garden Line	38
Riddles	36
Rollers	14, 48
Root Cutters	29
Roup Syringe	57
Rubber Sprinkler	40
Rustic Chairs, etc	54

S
Sack Truck	29
Sash	37
Saucers for Flower Pots	42
Sausage Stuffer	43
Saws	40
Sawing Machine	31
Scales	34, 43
Scythes, etc	40, 53
Scissors	40
Screens	36
Scraper—Road	23
" —Tree	41
Seed Sowers	17 to 23
" Drills	17 to 23
" Pans	42
Settees	54
Sieves	36

S
	PAGE
Sickles	53
" Grinder	25
Sheep Dip	61
Shovels	41
Shears—Lawn	53
" —Border	53
" —Grass	53
" —Hedge	53
" —Pruning	40
" —Lopping	40
" —Horse Clipping	40
" —Sheep	40
Smoke Puffer	67
Sod Cutter	49
Spades	41, 53
Spoon—Post Hole	41
Spraying Pumps, etc	64, 65
" Nozzles, etc	67
" Carts	36, 66
" Calendar	63
Sprinkling Carts	32, 52
Stands for Plants	42
Sun Dial	54
Sweeper—Lawn	48
Sweat Scraper	41
Stakes	55
Staples—Fencing	36
Syringes	66, 67

T
Tape Measure	41
Tedder	25
Tether Chain and Pin	38
Thermometers	41
" Dairy	45
" Incubator	57
Thrasher	31
Tomato Supports	36
Torch Asbestos	67
Transplanting Machine	18
" Trowel	41
Tree Tubs	55
" Scraper	41
Trellis	36, 54
Tree Guards	54
" Pruners	40
Trowels	41
Turf Edgers	49, 53
" Lifter	53

U
Umbrella	54

V
Vaporizer	64
" Bellows	66
Vases for Cut Flowers	42
Vegetable Slicer	43
" Graters	43
Veterinary Remedies	61

W
Wagons	32
Wagon Jacks	34
Watering Pots	41
" Carts	32, 52
Weeders—Horse	13
" —Hand	13, 23
" —Lawn	53
Weed Destroyer	56
Wheelbarrows	33, 49
Wheel Hoes	19 to 23
Wire Fencing	36
" Staples	36
" Stretcher	36
" Arches	54
" Trellis	36, 54
" Tree Guards	54
" Plant Stands	42
Worm Eradicator	56
Wrench	41

McIlroy & Emmet, 106 & 108 Liberty St., N. Y.

BOOKS ON HORTICULTURE AND AGRICULTURE

THE ACKNOWLEDGED AUTHORITIES

By Peter Henderson

FOR GARDEN, GREENHOUSE AND FARM.

GARDENING FOR PLEASURE.

Was written to meet the wants of those desiring information on gardening for private use. Its scope therefore embraces directions for the culture and propagation of Flowers, Vegetables and Fruits in the garden and under glass. It has had a large sale, and gone through several editions, the present having been revised and greatly enlarged by the author in 1888; it exhaustively treats on the Vegetable Garden, Flower Garden, Fruit Garden, Greenhouse, Grapery, Window Garden, Lawn, the Water Garden, etc.

404 Pages. Fully illustrated. Price, postpaid, $2.00.

GARDENING FOR PROFIT.

If you wish to grow Vegetables for Sale, read GARDENING FOR PROFIT. Written particularly for the Market Gardener and Truck Farmer, yet it is of equal value for large private gardens. The first edition of GARDENING FOR PROFIT was published in 1866; it has been revised twice since, and its sale has been so large that up to this time 43 editions have been printed. *The present edition* was revised and greatly enlarged in the summer of 1886. Its scope has been greatly extended since the earlier editions were published. The varieties in vegetables recommended for market culture have also been carefully revised; so that what is now advised to plant are kinds in general use at this time.

375 Pages. Fully illustrated. Price, postpaid, $2.00.

NEW HANDBOOK OF PLANTS AND GENERAL HORTICULTURE.

... ISSUED 1890. ...

The original HANDBOOK OF PLANTS was issued in 1881; in *this new edition* all the new genera of importance are added. All botanical terms are given, and also a very full list of popular names, and all the generic names are accentuated. The natural system of arrangement is adopted in the descriptions instead of the Linnæan or artificial system; and a carefully compiled glossary of the technical terms used in describing plants. A monthly calendar of operations for the greenhouse and window garden, flower, fruit and kitchen garden, will undoubtedly render this edition valuable as a book of reference. Very full instructions are given for the culture and forcing of all Fruits, Flowers and Vegetables of importance; in short, there is sufficient matter given on all gardening subjects to allow this book to be termed **The American Gardener's Dictionary**. And it will be better adapted to the wants of American horticulturists than any of the more costly foreign works on gardening, for, though from a foreign standpoint these are all they claim to be, yet for our American climate much of the information, and especially the gardening instructions, are not only useless, but actually misleading. *526 Pages. Profusely illustrated. Price, postpaid, $4.00.*

PRACTICAL FLORICULTURE.

Although written especially for the Commercial Florist, it is equally valuable for the amateur and all having conservatories, greenhouses, window gardens, etc. PRACTICAL FLORICULTURE was first issued in 1868, and has gone through many editions and had an enormous sale, and is admitted to be the leading American authority on this subject. This present edition of PRACTICAL FLORICULTURE was greatly enlarged by Mr. Henderson in 1887, and revised to keep abreast of the times, as there are now many superior methods of propagation and culture of flowers and plants—and many improved varieties of plants—all of which have been fully treated in this new edition.

325 Pages. Fully illustrated. Price, postpaid, $1.50.

GARDEN AND FARM TOPICS.

CONTENTS.—Popular Bulbs—Window Gardening—Plants in Rooms—Propagation of Plants by Cuttings, Layers, Divisions and Seeds—Rose Growing in Winter—Greenhouse Structures and Heating—Formation and Renovation of Lawns—Onion Growing for Market—How to Grow Cauliflower for Market—Growing and Preserving Celery—Strawberry Culture—Root Crops for Farm Stock—Culture of Alfalfa and Lucerne—Manures and their Application—Market Gardening around New York—The Use of the Feet in Seed Sowing and Planting—Draining.

244 Pages. Illustrated. Price, postpaid, $1.00.

HOW THE FARM PAYS. By PETER HENDERSON and WILLIAM CROZIER.

An acknowledged authority for Farmers. Gives all the Latest Methods of Growing Grass, Grain, Root Crops, Fruits, etc.; and all about Stock, Farm Machinery, etc., etc. It is written in a plain and easy-to-be-understood language. Everything pertaining to scientific or abstruse subjects has been ignored, the information given being the most direct *to make the work of the farm pay,* which the so-called scientific farmer rarely does. This is perhaps the first book of the kind ever written by two men while actually engaged in the work which, to both, has been such a continued success—hence, their advice is practical and doubly valuable. *400 Pages. Profusely illustrated. Price, postpaid, $2.50.*

BULB CULTURE.

CONTENTS.—Descriptions of Bulbs, alphabetically arranged, with Special Cultural Instructions for each—Winter Flowering Bulbs—Summer Flowering Bulbs—Forcing Bulbs. *24 Pages. Price, postpaid, 25 Cents.*

Condensed Vegetable and Flower Seed Culture.

An eight-page pamphlet, containing, in a condensed form, instructions for the cultivation of Garden Vegetables and Flowers from seeds. Also, full directions for making Hot-Beds and Cold Frames. *Price, 10 Cents.*

The Culture of Water Lilies and Aquatics.

Species and Varieties; their Culture in Natural and Artificial Ponds and Basins, Tubs, etc. *42 Pages. Illustrated. Price, postpaid, 25 Cents.*

Insects and Plant Diseases, with Remedies.

Insecticides and Fungicides: How to Mix and Apply Them—Injurious Insects, with Remedies—Plant Diseases (such as Mildew, Rust, Rot, etc., etc.), with Remedies. *36 Pages. Illustrated. Price, postpaid, 25 Cents.*

Special Offer. *It ordered at one time, we will supply the full set of ten books offered on this page, carriage prepaid, for $10.00. (Separately, they would cost $13.85.) This set of books forms A COMPLETE LIBRARY OF THE GARDEN, GREENHOUSE AND FARM.* **for $10.00**

BOOKS ON HORTICULTURE, AGRICULTURE AND KINDRED SUBJECTS

Delivered Free in the U. S. at these prices by Peter Henderson & Co., New York.

PLANTS, FLOWERS AND FLOWER GARDENING.

Gardening for Pleasure. By PETER HENDERSON. *(See description, page 1.)* $2.00

Home Floriculture. (The Cultivation of Garden and House Plants.) By REXFORD. Written particularly for amateurs..........(Illustrated) 1.50

The English Flower Garden. By W. ROBINSON. (Imported book.) Position, arrangement, with best plants for various purposes, and their culture..(Illustrated) 6.00

Gardening for Ladies. By MRS. J. C. LOUDON. 2d American Edition.......... 1.50

The Wild Garden. By ALFRED PARSONS. (Imported book.) The natural grouping of hardy plants; the best for various effects, culture, etc..........(Illustrated) 5.00

Sub-Tropical Gardening. By WILLIAM ROBINSON. (Imported.) The best plants for this purpose and arrangement..........(Illustrated) 2.50

The Bamboo Garden. By A. B. F. MITFORD. Varieties, descriptions, arrangement, culture of sorts adapted to American climates..........(Illustrated) 3.00

The Water Garden. By WM. TRICKER. All water plants described; how to grow in tubs, ponds, etc.; the formation of artificial ponds, utilization of natural; propagation, culture, wintering, etc.......... 2.00

The Old-Fashioned Garden and Hardy Perennials. By J. WOOD. (Imported.) Old-fashioned flowering and foliage plants; shrubberies, borders, etc..(Illustrated) 2.00

Garden Making. By PROF. BAILEY. Instructions for beginners and a book of reference for the skilled gardener. Covers the whole subject, laying out and planting small city yards and large suburban grounds, plants, trees, bedding, pruning, vegetables, fruits, scientific truths in simple language..........(Illustrated) 1.00

GREENHOUSE AND WINDOW GARDENING.

Gardening for Pleasure. By PETER HENDERSON. *(See description, page 1.)* $2.00

Practical Floriculture. By PETER HENDERSON. *(See description, page 1.)* 1.50

Cut Flowers and How to Grow. By M. A. HUNT. The practical cultivation of Roses and other flowers for cut flowers, by an authority..........(Illustrated) 2.00

Greenhouse and Stove Plants. By T. BAINS. (Imported.) A masterly English work by their foremost authority..........(Illustrated) 3.50

Greenhouse Management for Amateurs. By W. J. MAY. (Imported.) How to build and heat greenhouses and frames; suitable plants and culture. (Illustrated) 2.00

Window and Parlor Gardening. By N. JÖNSSON-ROSE. The daily care of house plants and allied subjects; a book of reference for the amateur..........(Illustrated) 1.25

The Window Flower Garden. By J. J. HEINRICH. The personal experience of a practical florist..........(Illustrated) .75

House Plants; How to Succeed with Them. By LIZZIE P. HILLHOUSE. For women who grow house plants, by a woman who has success..........(Illustrated) 1.00

House Plants as Sanitary Agents. By DR. ANDERS. Relations of vegetation in floriculture, forests, plantations, etc., to health and disease.......... (Illustrated) 1.50

Greenhouse Construction, also Hot-beds and Frames. By PROF. L. R. TAFT. All details for florists and amateurs; heating and ventilating..........(Illustrated) 1.50

Greenhouse Heating by Hot Water and Steam. By A. B. FOWLER. Best systems, location, selecting apparatus, computing and other details..........(Illustrated) .75

CULTURES OF SPECIAL PLANTS.

Amaryllideæ (*Amaryllis Family*), including Alstromerias and Agaves. By J. G. BAKER, of Kew Gardens, England. (Imported.)..........(Illustrated) $2.00

Azalea Culture. By R. J. HALLIDAY. A practical treatise on propagation and cultivation of Azalea Indica..........(Illustrated) 2.00

Begonia Culture. By B. C. RAVENSCROFT. (Imported.) Under glass and open air; directions for both amateurs and professionals.......... (Illustrated) .50

Begonias, Tuberous. By several practical growers. Gives cultural directions and general management.......... .25

Bulb Culture. By PETER HENDERSON. *(See description, page 1.)* .... .25

Bulb Culture, Popular. By W. D. DRURY. (Imported.) An English guide to cultivation of bulbs under glass and open air..........(Illustrated) .50

Bulbs and Tuberous-Rooted Plants. By C. L. ALLEN. Descriptions, propagation, culture in dwelling, greenhouse and garden. A splendid work..........(Illustrated) 2.00

Cactus Culture for Amateurs. By W. WATSON. (Imported.) Descriptions and full cultural instructions..........(Illustrated) 2.00

Camellia Culture. By R. J. HALLIDAY. Their practical cultivation and propagation..........(Illustrated) 2.00

Carnation Culture, American. By L. L. LAMBORN. Varieties, classification, propagation, culture. A practical work..........(Illustrated) 1.50

Carnation Culture for Amateurs. An English work, by B. C. RAVENSCROFT. Carnations and Picotees of all classes; pots and open ground culture..(Illustrated) .50

Chrysanthemum Culture for America. By JAMES MORTON. A thorough work, fully covering the subject for America..........(Illustrated) 1.00

Chrysanthemum Culture for Amateurs and Professionals. An English work, by B. C. RAVENSCROFT. Cultivation for both exhibition and market......(Illustrated) .50

Chrysanthemum, Growth of the Plant. By EDWIN MOLYNEUX. A practical English work on culture, etc..........(Illustrated) .50

Dahlia, The. By L. K. PEACOCK. New and valuable work. Classes, varieties, descriptions, cultivation, history..........(Illustrated) .50

Ferns and Ferneries...........(Illustrated) .25

Ferns in their Homes and Ours. By PROF. J. ROBINSON. Our native Ferns, when and where to find them; how to grow them at home..........(Illustrated) 1.50

Ferns and Fern Culture. By J. BIRKENHEAD. (Imported.) Selections and culture for cold, warm and stove ferneries, Wardian cases, dwellings, etc. (Illustrated) .85

Ferns, The Book of Choice. By GEO. SCHNEIDER. A beautiful work in three volumes; the best ferns, descriptions, cultures, etc.......... (Elegantly illustrated) 18.00

Irises, Bulbous. By PROF. MICHAEL FOSTER. (Imported.) Species, varieties, descriptions, time of flowering, habitat and culture for each..........(Illustrated) 2.00

Lilies and their Culture. By DR. WALLACE. (Imported.) All varieties are described; their native habitats and conditions, culture..........(Illustrated) 1.50

Orchids. The Amateur Cultivator's Guide Book. By H. A. BURBERRY, Orchid grower to Rt. Hon. Joseph Chamberlain. Varieties, descriptions and how to grow in cool, intermediate and warm houses. (Imported.)..........(Illustrated) 2.00

Orchid Grower's Manual. By B. S. WILLIAMS. (Imported.) Descriptions of 2,500 species and varieties, culture and other information..........(Illustrated) 10.00

Orchids, Their Culture and Management. New Edition. By W. WATSON, Royal Gardens, Kew, England. (Imported.) Descriptions of all kinds in general cultivation. Elegant illustrations and colored plates.......... 10.00

Pansy, The. By J. SIMKINS. (Imported.) Tells all about them; how to grow and how to show them..........(Illustrated) .85

Rhododendrons. By E. S. RAND, JR. Revised Edition..........(Illustrated) 1.50

Rose, The. By H. B. ELLWANGER. Revised Edition. Varieties, classification, characteristics, cultivation, pruning, propagation, etc..........(Illustrated) 1.25

Rose, Parsons on the. By S. B. PARSONS. Revised Edition. Propagation, culture, training, classification and descriptions..........(Illustrated) 1.00

Rose Culture, Secrets of. By W. J. HATTON, florist. Paper. Rose houses, heating, management; best Roses for all purposes, etc..........(Illustrated) .50

Roses, A Book about. By DEAN S. REYNOLDS HOLE. 14th Edition. (Imported.) A reliable English guide to Rose culture.......... 1.35

Rose Book, The Amateur's. By SHIRLEY HIBBARD. (Imported.) Cultivation under glass and in the garden; formation of rosarium, etc..........(Illustrated) 2.75

Rose Garden. By WM. PAUL. A valuable work by an English specialist; descriptions, culture, etc..........(Beautifully illustrated and 20 colored plates) 10.00

Roses in Pots, The Cultivation of. By WM. PAUL. (Imported.)..........(Illustrated) .85

Sweet Peas. By REV. W. T. HUTCHINS. Varieties, cultivation, etc..........(Illustrated) .20

HARDY GARDENING AND LANDSCAPING.

Handbook of Practical Landscape Gardening. By F. R. ELLIOTT. Designs for small city lots and large suburban grounds..........(Illustrated) $1.50

Landscape Gardening, or How to Lay out a Garden. By EDWARD KEMP. A guide in choosing, forming or improving small places and large estates......(Illustrated) 2.50

Lawns and Gardens. By N. JÖNSSON-ROSE. Practical landscaping. How to beautify home surroundings; plans, best plants and their places..........(Illustrated) 3.50

The Royal Parks and Gardens of London. By NATHAN COLE. (Imported book.) Their embellishment, flower beds and borders, sub-tropical bedding, geometrical designs, the plants used and their propagation and culture..........(Illustrated) 1.25

The Parks and Gardens of Paris. By WM. ROBINSON. (Imported.) Considered in relation to the wants of other cities. 350 illustrations.......... 10.00

God's Acre Beautiful, or The Cemeteries of the Future. By WM. ROBINSON. (Imported book.)..........(Illustrated) 3.00

ENCYCLOPEDIAS OF PLANTS AND THEIR CULTURE.

Handbook of Plants. By PETER HENDERSON. Descriptions and American culture of species. *(See description, page 1.)* $4.00

Dictionary of Gardening. An English work by GEO. NICHOLSON, A. L. S. Botanical classification, full descriptions of both species and varieties, with cultural directions; practical, useful and valuable. 4 vols..........(Illustrated) 20.00

PLANT BREEDING, PROPAGATION, PRUNING.

Plant Breeding. By PROF. L. H. BAILEY. The philosophy of crossing, variation and improvement of plants..........(Illustrated) $1.00

Complete Guide to the Multiplication of Plants. (The Nursery Book.) By PROF. BAILEY. Full directions from seed, layers, cuttings, grafts, bulbs, etc.... 1.00

The Propagation of Plants. By A. S. FULLER. Hybridizing, crossing; modes of propagation and multiplication..........(Illustrated) 1.50

The Pruning Book. By PROF. BAILEY. Issued 1897. Where, how and when to prune fruit and ornamental trees and bushes..........(Illustrated) 1.00

The Horticulturist's Rule Book. By PROF. L. H. BAILEY. 4th Edition. Full of useful information for fruit growers, truck gardeners, florists and others.......... .75

PLANT LIFE, HISTORY, PHYSIOLOGY, ETC.

New American Botanist and Florist. By ALPHONSO WOOD. Revised by O. R. WILLIS, Ph.D. Lessons in structure, life and growth of plants and analytical descriptions of nearly 4,500 species of plants of eastern United States.......... $2.00

The Geological History of Plants. By SIR J. WILLIAM DAWSON.......... 1.75

Origin of Cultivated Plants. By ALPHONSE DE CANDOLLE.......... 2.00

Trees, Plants and Flowers; Where and How they Grow. A familiar history of the vegetable kingdom. By W. L. BAILEY.......... .75

Popular Treatise on the Physiology of Plants. By DR. P. SORAUER. Scientific, but specially written for gardeners who wish to understand the principles of plant construction, absorption and assimilation of food, etc......(Illustrated) 3.00

Modification of Plants by Climate. By A. A. CROZIER. How it influences form, color, fruitfulness, etc.......... .25

The Wonders of Plant Life. By MRS. S. B. HERRICK..........(Illustrated) 1.50

How Plants Grow. By DR. ASA GRAY. Why plants grow, etc.......... 1.00

How Plants Behave. By DR. ASA GRAY. How they move, climb, etc.......... .75

Evolution of Plant Life from the Lower Forms. By G. MASSEE.......... 1.00

Effects of Cross and Self Fertilization. By CHAS. DARWIN.......... 2.00

On the Colors of Flowers. By G. ALLEN. An evolutionary argument.......... 1.00

BOTANIES.

Lessons and Manual of Botany. By DR. ASA GRAY..........(Illustrated) $2.50

Field Botany. By W. P. MANTON. A handbook for the collector.......... .50

Botany, Field, Forest and Garden. By DR. ASA GRAY. Revised by PROF. L. H. BAILEY. A simple guide for gardeners and amateurs to study the structures and names of the commoner plants east of the Mississippi..........(Illustrated) 1.75

The Microscope in Botany. Translated from German of DR. BEHRENS. A guide for the microscopical investigations of vegetable substances........(Illustrated) 5.00

AMERICAN FLORA (Not Cultural).

HOW TO KNOW OUR WILD FLOWERS, FERNS, MOSSES, ETC.

Familiar Flowers of Field and Garden. By F. SCUYLER MATHEWS. Descriptions. 200 illustrations, rendering identification easy, especially so by the aid of arrangement into color classes and seasons of bloom.......... $2.25

How to Know Wild Flowers. By MRS. DANA. A guide to haunts and habits; their identification easy for amateurs..........(Illustrated) 1.75

Wild Flowers of North America. By PROF. GOODALE, of Harvard Botanic Gardens. 51 colored plates and numerous other illustrations.......... 10.00

The Wild Flowers of the Northeastern States. By ELLEN MILLER and MARGARET C. WHITING. Careful, easily understood descriptions, aided by illustrations (life-size), enable any one to identify and name our wild flowers.......... 4.50

Flora of the Southern States. By CHAPMAN. Third Edition.......... 4.00

Flora of North America. By ASA GRAY and SERENO WATSON.......... 2.75

Ferns and Evergreens of New England. By EDWARD KNOBEL. A guide to the identification of native ferns and mosses..........(Illustrated) .75

Manual of Mosses of North America. By MESSRS. LESQUEREUX and JAMES. The latest and recognized authority describing all of our species..........(Illustrated) 4.00

The Orchids of New England. By HENRY BALDWIN. Illustrated from nature...... 1.50

The Lady's Slippers—Cypripediums—of Northern and Eastern United States. By C. S. YOUNG. All species..........(Illustrated) .75

Hepaticæ of North America. By PROF. UNDERWOOD. The latest and best treatise. 1.50

A Guide to the Study of Lichens. By ALBERT SCHNEIDER, M.D., Ph.D. (Illustrated) 2.50

Medicinal Plants of the United States. By DR. C. F. MILLSPAUGH. Description, locality, history, properties, preparation and effects of over 1,000 plants. 180 colored pages; 1,000 other illustrations; 2 vols.......... 42.00

Our Edible Toad Stools and Mushrooms. How to distinguish the edible from the poisonous. 30 colored plates and numerous other illustrations.......... 7.50

PESTS—INSECTS, PLANT DISEASES, WEEDS.

Economic Entomology. By PROF. SMITH, one of the highest authorities. Insects easily identified; preventives, machinery, fungous diseases, etc.......(Illustrated) $2.50

Insects Injurious to Fruits. By PROF. SAUNDERS. Insects that prey on trees and best methods of destroying them..........(Illustrated) 2.00

Insects and Insecticides. By PROF. WEED. Tells how to combat insects in field, orchard, garden, greenhouse and dwelling..........(Illustrated) 1.50

Fungi and Fungicides. By PROF. WEED. Fungous diseases of plants, etc., and their treatment..........(Illustrated) 1.00

The Spraying of Plants. By PROF. LODEMAN. Insects and fungi; liquids and powders; application and apparatus; complete and reliable..........(Illustrated) 1.00

Diseases of Field and Garden Crops. By W. G. SMITH. An English work for gardeners and farmers; description, cause, averting and combating. (Illustrated) 1.50

Weeds and How to Eradicate Them. By PROF. THOS. SHAW..........(Illustrated) .75

We will procure for our customers any book published, if still in print, and deliver free in the U. S. at the regular retail price.

BOOKS ON HORTICULTURE, AGRICULTURE AND KINDRED SUBJECTS Continued.

Delivered Free in the U. S. at these prices by Peter Henderson & Co., New York.

VEGETABLE GARDENING.

Gardening for Pleasure. By PETER HENDERSON. *(See description, page 1.)* $2.00

Garden Making. By PROF. BAILEY. *(See description under "Plants and Flower Gardening" section.)* 1.00

Vegetable Gardening. By PROF. S. B. GREEN, Professor of Horticulture, University of Wisconsin. A new and original work. Full of practical information about the growing of vegetables for both home use and for marketing..........(Illustrated) 1.25

Vegetable Garden, The. Translated by W. ROBINSON from the French of H. DE VILMORIN. An exhaustive work on vegetables for cool and temperate climates; descriptions of types, varieties, cultivation and other valuable information. An indispensable reference book........ 5.00

The Forcing Book. Winter vegetables. *(See description under "Market Gardening" section.)* 1.00

MARKET GARDENING AND TRUCK FARMING.

Gardening for Profit. By PETER HENDERSON. *(See description, page 1.)* $2.00

Truck Farming at the South. By DR. A. OEMLER. A guide to raising vegetables for northern markets, culture, packing, etc., by an experienced and successful grower........(Illustrated) 1.50

Vegetable Growing in the South for Northern Markets. By PROF. ROLFS, of Florida Agricultural College. Practical and valuable information.....(Illustrated) 1.25

Success in Market Gardening. By W. W. RAWSON. Vegetables out-of-doors and under glass. Specially adapted to New England climate........(Illustrated) 1.00

The Young Market Gardener. By T. GREINER. A guide to beginners in market vegetables; outside culture, hot-beds, frames, preparing and selling, etc....(Illustrated) .50

The Forcing Book. By PROF. BAILEY. The cultivation of winter vegetables in glass houses. The best and most complete book on this subject. Invaluable either for those who grow for home consumption or for market........(Illustrated) 1.00

CULTURES OF SPECIAL VEGETABLES.

Asparagus Culture. By JAS. BARNES and WM. ROBINSON. (Imported.) The best methods employed in England and France........(Illustrated) $0.50

Cabbages, How to Grow. By J. J. H. GREGORY. Details of culture, keeping, marketing, etc........(Illustrated) .50

Cabbages and Cauliflowers for Profit. By J. M. LUPTON. A new book on this subject by a successful grower........(Illustrated) .50

Carrots and Mangels. By J. J. H. GREGORY. How to raise them, keep them and feed them........(Illustrated) .30

Celery, Kalamazoo Culture of. By G. VON BOCHOVE. Improved methods of culture. "The Secret of Success," and full information........(Illustrated) .50

Mushrooms, How to Grow Them. By WM. FALCONER. The best and most practical American work on growing for home use or for market, by a successful specialist........(Illustrated) 1.50

Mushroom Culture. By W. ROBINSON. (Imported.) England's standard authority on this subject........(Illustrated) .50

Mushroom Culture for Amateurs. By W. J. MAY. (Imported.) An English work, giving methods of growing in houses, sheds, cellars, shelves and out-of-doors...... .50

Onion Culture, The New. By T. GREINER. For the home garden or market; new and highly valuable methods are described........(Illustrated) .50

Potato Culture, The A B C of. By W. B. TERRY. How to grow quantity and quality, and other new and valuable information........(Illustrated) .40

Potato Culture, The New. By E. S. CARMAN. New and profitable methods; trench system, etc. Results of 15 years' experiments........(Illustrated) .75

Sweet Potato Culture. By JAMES FITZ. Full instructions from starting the plants to harvesting and storing; the Chinese Yam, etc........(Illustrated) .60

Rhubarb Culture. By F. S. THOMPSON. A complete guide by one of the largest practical growers........(Illustrated) 1.00

Squashes. By J. J. H. GREGORY. Soil selection and preparation, culture, gathering, winter storing, etc........(Illustrated) .00

The Tomato. By W. IGGULDEN, F. R. H. S. (Imported.) The English method of maintaining under glass a continuous supply........(Illustrated) .60

Tomato Culture. By DAY, CUMMINS and A. I. ROOT. Culture in field, under glass and in the South; for home, for market, for canning factories........(Illustrated) .40

GENERAL FRUIT AND NUT CULTURE.

The Principles of Fruit Growing. By PROF. BAILEY. A new work and one of the most valuable ever written on the subject, science and practice....(Illustrated) $1.25

American Fruit Culturist. By J. J. THOMAS. 20th Edition; just revised and enlarged. A handbook of everything pertaining to fruit culture........(Illustrated) 2.50

The Fruit Garden. By P. BARRY. A standard work on fruit culture by an experienced author and nurseryman........(Illustrated) 2.00

Fruit Culture. By W. C. STRONG, Vice-President American Pomological Society. New and Revised Edition. Making it the latest work on the subject. (Illustrated) 1.00

Fruits and Fruit Trees of America. By A. J. DOWNING. Culture, propagation and management, with descriptions and illustrations of native and foreign fruits...... 5.00

The Practical Fruit Grower. By C. T. MAYNARD. Just what the beginner needs and the successful man practices........(Illustrated) .50

Small Fruit Culturist. By A. S. FULLER. Re-written, enlarged and up to date; propagation, culture, varieties, marketing, etc........(Illustrated) 1.50

The Orchard House. By J. R. PEARSON. The English method of growing fruits in glass houses. Construction and management........ .85

The Nut Culturist. By A. S. FULLER. Propagation, cultivation, marketing of nut-bearing trees and shrubs........(Illustrated) 1.50

Nuts for Profit. By J. R. PARRY. Germination, budding, grafting, cultivation, harvesting, marketing, receipts for preparation and serving........(Illustrated) 1.00

CULTURES OF SPECIAL FRUITS.

Apple Culture, Field Notes on. By PROF. BAILEY. Practical and valuable instruction from planting to harvesting........(Illustrated) $0.75

Berry Book, The Biggle. A handy small work on berries, particularly strawberries........(Illustrated) .50

California Fruits and How to Grow Them. By E. J. WICKSON. Methods and experience of growers; varieties for certain districts, etc........(Illustrated) 3.00

Cider Maker's Handbook. By J. M. TROWBRIDGE. Making and keeping in perfection, based on scientific facts........(Illustrated) 1.00

Cranberry Culture. By J. J. WHITE. Location, preparation, planting, management, picking, keeping, etc., etc........(Illustrated) 1.25

Cape Cod Cranberries. By JAS. WEBB. A valuable handbook by a successful cultivator........(Illustrated) .40

Florida Fruits and How to Raise Them. By H. HARCOURT. Cultivation, management, marketing of all fruits adapted to semi-tropical regions of the U. S.; evaporating fruits and how to use them........(Illustrated) 1.25

Grape Culturist. By A. S. FULLER. One of the very best works on cultivation and management of hardy grapes........(Illustrated) 1.50

Grape Growing and Wine Making, American. By PROF. G. HUSMANN. Revised 1895. Garden and vineyard management from planting to harvesting, both in the East, West and in California; all about making wine........(Illustrated) 1.50

Grape Grower's Guide (under Glass). By WM. CHORLTON. Cultivation suited to America in warm and cold graperies, construction, heating, etc........(Illustrated) .75

Grape Training, American. By PROF. BAILEY. A new book, illustrating and describing all practical systems in detail........ .75

CULTURES OF SPECIAL FRUITS—Continued.

Orange Culture (in Florida, Louisiana and California). By REV. T. W. MOORE. Oranges, lemons and limes; culture, gathering, packing, shipping, etc..(Illustrated) $1.00

The Olive. Its Culture in Practice and Theory. By A. T. MARVIN. Reliable treatment of all phases of the subject........(Illustrated) 2.00

Peach Culture. By HON. J. A. FULTON. Revised. The best work on growing peaches for profit or home use........(Illustrated) 1.50

Pear Culture for Profit. By P. T. QUINN. Soils, preparation, planting, management, harvesting, marketing........(Illustrated) 1.00

Quince Culture. By W. W. MEECH. Revised and enlarged. Varieties, propagation, cultivation, diseases, insects and remedies........(Illustrated) 1.00

Strawberry Culturist. By A. S. FULLER. Field, garden, forcing and pot culture; hybridizing, varieties, etc........(Illustrated) .25

Strawberry Culture, The A B C of. By T. B. TERRY. The latest on this subject and by an experienced grower........(Illustrated) .40

SHADE TREES, FORESTRY AND TIMBER.

Trees for Street and Shade. By MESSRS. POWELL and MCMILLAN. From nursery to permanent location; what, where and how to plant trees for city streets........ $0.25

Practical Forestry. By A. S. FULLER. Varieties, propagation, planting and cultivation of both evergreen and deciduous........(Illustrated) 1.50

Forest Planting. By H. N. JARCHOW, LL.D. Issued 1897. Restoration, maintenance and care of wood and timber lands on plains and mountains........(Illustrated) 1.50

Timber and Some of its Diseases. By H. M. WARD. Of value to every one interested in the care of trees........(Illustrated) 1.75

Outlines of Forestry. By E. J. HOUSTON, A. M., of the Pennsylvania Forestry Association. The principles underlying the science; the effects on climate, country and rainfall; how to make amends and prevent further loss........ 1.00

Elements of Forestry. By F. B. HOUGH, Ph.D., Chief of Forestry Division, U. S. Department of Agriculture. Planting and care for both profit and ornament; creation and care of woodlands, etc........(Illustrated) 2.00

The Woods of the United States. By PROF. C. S. SARGENT. Structure, qualities, uses, with geographical and other notes on the trees that produce them........ 1.00

Lumber and Log Book. By J. L. SCRIBNER. New and Enlarged Edition. Quick computation of measurement, weight, etc., of lumber in all forms, etc., etc...... .25

Maple Sugar and the Sugar Bush. By PROF. COOK. How to make maple sugar; new apparatus, etc........(Illustrated) .35

AMERICAN SYLVA (Not Cultural).

HOW TO KNOW NATIVE TREES, SHRUBS, ETC.

Familiar Trees and their Leaves. By F. SCUYLER MATHEWS. Characteristics, descriptions and illustrations of over 200 types, common and exceptional........ $1.75

The Trees of Northeastern America. By CHAS. S. NEWHALL. The descriptions and illustrations enable any one to identify and name them........ 2.50

The Shrubs of Northeastern America. By CHAS. S. NEWHALL........ 2.50

Trees and Shrubs of New England. By EDWARD KNOBEL. A guide to find the names of all wild species by their leaves........(Illustrated) .75

Trees of the Northern United States (east of the Rocky Mountains). By PROF. APGAR. Their study, description and determination........(Illustrated) 1.25

GENERAL AGRICULTURE AND FARMING.

How the Farm Pays. By MESSRS. HENDERSON and CROZIER. *(See page 1.)* $2.50

American Farm Book. By R. L. and L. F. ALLEN. Revised. A compendium of farming in all of its details; one of the best works on the subject.....(Illustrated) 2.50

Our Farming. By T. B. TERRY. The experience of 20 years' successful and up-to-date farming; valuable for reference; no farmer should be without it. (Illustrated) 2.00

A Handbook for Farmers and Dairymen. By F. W. WOLL, Professor of Agricultural Chemistry, University of Wisconsin. A book of reference, of great value, facts, tables, formulas, receipts, cultivation of crops, feeding animals, etc; revised and brought up to date 1897........(Illustrated) 1.50

Manual of Agriculture. By MESSRS. EMERSON and FLINT. A new edition, revised by DR. GOESSMANN, Professor of Chemistry, Massachusetts Agricultural College... 1.50

Book of the Farm. By GEO. E. WARRING, JR. Buying, leasing, fences, buildings, implements, drainage, subsoiling, rotation, etc., etc........(Illustrated) 2.00

Agriculture. By R. HEDGES WALLACE, Professor Victoria (Australia) Department of Agriculture. The principles of modern agriculture, resulting from scientific investigations founded on natural laws, applicable to any agricultural country. 1.25

A LITTLE SCIENCE RELATED TO FARMING.

Agriculture in Some of its Relations with Chemistry. By F. H. STORER, Professor of Agricultural Chemistry, Harvard University. 3 vols. New Edition, with important revisions. Comprehensive treatment of hundreds of subjects; an exhaustive work of great value to the farmer........(Illustrated) $6.00

How Crops Grow. By PROF. SAMUEL JOHNSON, of Yale College. Agricultural plants, composition, development, requirements, tables of analysis, etc.; indispensable to farmers who want to understand the "whys and wherefores." (Illustrated) 2.00

How Crops Feed. By PROF. SAMUEL JOHNSON. Scientific facts of atmosphere and soil as related to nutrition of plants, etc........(Illustrated) 2.00

Practical Farm Chemistry. By T. GREINER. A handbook of profitable crop feeding........(Illustrated) 1.00

Chemistry of the Farm. By R. WARRINGTON, F.C.S. The relations of chemistry and agriculture, clear and concise; of great value to all tillers of the soil........ 1.00

The Science of Agriculture. By JAS. LLOYD. An English work on soils, plant food, manures, rotation of crops, pasturage, management of live stock, dairy, etc....... 4.00

CULTURES OF SPECIAL FARM CROPS.

Broom Corn and Brooms. By Editors of the "AMERICAN AGRICULTURIST." Raising broom corn and making brooms on large or small scale........(Illustrated) $0.50

Corn Culture (Indian). By C. S. PLUMB, Director Indiana Experiment Station. Practical as well as scientific instructions, covering all details........(Illustrated) 1.00

Flax Culture. By several experienced growers. Selecting and preparing ground; culture, harvesting and marketing........ .30

Grasses and Forage Plants. By CHAS. L. FLINT. New Edition. Varieties, nutritive values, culture, curing, management grass land, etc........(Illustrated) 2.00

Grasses and Clovers. Field Roots. Forage and Fodder Plants. By PROF. THOS. SHAW. Food values, cultivation, etc........(Illustrated) .50

Grasses of North America. By PROF. W. J. BEAL, of Michigan Agricultural College. Descriptions, structure, form, development, directions for cultivation under varied conditions; in 2 vols. In preparation........(Illustrated) ....

Peanut Plant. Its Cultivation and Uses. By B. W. JONES. Instructs the beginner how to raise good crops........(Illustrated) .50

Sugar Beet, The. By L. S. WARE. Varieties, soils, tillage, harvesting; the industry in Europe, etc........(Illustrated) 4.00

Silage, Ensilage and Silos. By MANLY MILES. Practical treatise on ensilage of fodder corn, etc........(Illustrated) .50

Sorghum. By PETER COLLIER, Ph. D., (late chemist U. S. Department of Agriculture.) Culture and manufacture as a source of sugar, syrup and fodder......(Illustrated) 3.00

Tobacco Culture. Full practical details by fourteen experienced growers in different sections of the country........(Illustrated) .25

Tobacco Leaf. By KILLE, DREW and MYRICK. Issued 1897. Approved methods of culture, harvesting, curing, packing, selling and manufacturing. Every process in field, barn and factory made plain........(Illustrated) 2.00

Wheat Culture. By D. S. CURTISS. How to double the yield, varieties, improved machinery, etc........(Illustrated) .50

We will procure for our customers any book published, if still in print, and deliver free in the U. S. at the regular retail price.

BOOKS ON HORTICULTURE, AGRICULTURE AND KINDRED SUBJECTS | Continued.

Delivered Free in the U. S. at these prices by Peter Henderson & Co., New York.

SOILS AND MANURES.

The Soil. By F. H. KING, Professor Agricultural Physics, University of Wisconsin. Its nature, composition, functions, relations to plant life and principles of management; a distinct advance on the subject ....(Illustrated) $0.75

The Fertility of the Land. By PROF. ROBERTS, Director Cornell Agricultural Experiment Station. A valuable book to every tiller of the soil; the philosophy of controlling and increasing fertility through management of soil, water, rotation. 1.00

Talks on Manures. By JOSEPH HARRIS, M. S. Familiar talks on the whole subject of manures and fertilizers ....(Illustrated) 1.75

A Treatise on Manures. By DR. A. B. GRIFFITHS. (Imported.) A handbook of information on manuring, fertilizers and fertilizing substances ....(Illustrated) 3.50

Farming with Green Manures. By DR. C. HARLAN. 4th and Enlarged Edition. The advantage of soiling and green manuring; details of practice and effects. 1.00

Scientific Examination of Soils. Translated from the German of DR. WAHNSCHAFFE. Select methods of chemical analysis and physical investigations ....(Illustrated) 1.50

Rocks and Soils. By PROF. STOCKBRIDGE, President North Dakota Agricultural College. Origin, composition, characteristics, geological and agricultural; all about the soil on your farm, and what every farmer should understand ....(Illustrated) 2.50

DRAINAGE AND IRRIGATION.

Tile Drainage. By W. J. CHAMBERLAIN. The experience of forty years by a practical agriculturist who has with his own hands laid 15 miles of tiles ....(Illustrated) $0.40

Land Draining. By MANLY MILES. A handbook of principles, practice and construction of tile drains; what errors to avoid ....(Illustrated) 1.00

Irrigation Farming. By LUTE WILCOX. The application of water in the production of crops, appliances, principles and advantages ....(Illustrated) 2.00

Irrigation for Farm, Garden and Orchard. By HENRY STEWART. Methods and management to secure water for critical periods ....(Illustrated) 1.50

FARM BUILDINGS AND CONVENIENCES.

Barn Plans and Outbuildings. Ideas, suggestions, plans for barns, granaries, smoke, ice, poultry, dog, bird houses, etc.; rootpits, etc ....(Illustrated) $1.50

Farm Conveniences. What to do and how to do it in all departments of farm labor; home-made aids to farm work ....(200 engravings) 1.50

American Architecture, or Every Man a Complete Builder. Instructions and plans for cottages, houses, barns, stables, etc ....(Illustrated) 1.00

Ice Crop; how to Harvest, Ship and Store. By T. L. MILES. Ice houses, cutting, storing, shipping, tools, etc ....(Illustrated) 1.00

Land Measurer for Farmers. By PEDDER. Shows at once the contents of any piece of land, with various other useful tables for farmers .... .60

DAIRYING AND DAIRY FARMING.

Principles of Modern Dairy Practice. By G. GROTENFELT, President Finland Agricultural College; American edition authorized by F. W. Woll, Professor Agricultural Chemistry, University of Wisconsin. Bacteria and their relations to new methods of dairying, from the udder to butter and cheese ....(Illustrated) $2.00

Milk and its Products. By H. H. WING, Professor Dairy Husbandry, Cornell University. A new book, covering the whole field ....(Illustrated) 1.00

Dairying for Profit; or, the Poor Man's Cow. By MRS. M. E. JONES, Judge of Dairy Products at the World's Fair, Chicago, 1893. Should be in the hands of every one having anything to do with dairying. Cloth .... .50

A B C in Cheese Making. By J. H. MONRAD. Home cheese making; Chedder, French cream, Neufchatel and skim milk cheese ....(Illustrated) .50

Butter and Butter Making. By W. F. HAZARD. Producing and marketing .... .25

Milch Cows and Dairy Farming. By CHAS. L. FLINT. Breeds, breeding and management in health and disease; selection and forage, culture of ....(Illustrated) 2.00

Dairyman's Manual. By HENRY STEWART. A trustworthy handbook, covering the entire subject, with latest approved methods ....(Illustrated) 2.00

American Dairying. By H. B. GURLER. Herd, feed, management, marketing, modern appliances; private and creamery dairying ....(Illustrated) 1.00

FARM ANIMALS AND LIVE STOCK.

The Breeds of Live Stock. By J. H. SANDERS. Descriptions of all breeds of horses, cattle, sheep and swine; the principles of breeding ....(Illustrated) $3.00

Stock Breeding. By MANLY MILES, M.D. The application of the laws of development, heredity and improvement in breeding domestic animals ....(Illustrated) 1.50

Horses, Cattle, Sheep and Swine. By GEO. W. CURTIS. History, description, merits of different breeds; hints on selection and management; methods of breeders 2.00

Diseases of Horses and Cattle. By DR. D. MCINTOSH, V. S. Modern treatment of animal diseases; for the farmer and stockman; by an eminent veterinarian 1.75

Farmers' Veterinary Adviser. By PROF. JAS. LAW. A guide to prevention of disease in domestic animals as well as remedies and treatment ....(Illustrated) 3.00

American Cattle, Sheep and Swine Doctor. By PROF. GEO. H. DADD. (Illustrated) 1.50

Feeding Animals. By E. W. STEWART. The laws of animal growth applied to the feeding and rearing of horses, cattle, cows, sheep and swine ....(Illustrated) 2.00

Chart of Age of Domestic Animals. By A. LIAUTARD, V. S. Enables one to accurately determine age of horses, cattle, sheep, pigs and dogs ....(Illustrated) .50

Animal Castration. By A. LIAUTARD, V. S. A practical treatise on the castration of domestic animals ....(Illustrated) 2.00

HORSES AND MULES.

The Exterior of the Horse. Translated from the French of MESSRS. GOUBAUX and BARRIER, and edited by S. J. J. HARGER, V. M. D., Professor of Veterinary, University of Penna. A new and exhaustive treatise of unusual merit. Every horseman desiring the highest authority should have a copy. 346 illustrations and 34 plates. By G. NICOLET, Librarian, Veterinary School at Alfort $6.00

Horse Breeding. By J. H. SANDERS. The principles of heredity, selection, breeding, management; treatment of diseases peculiar to breeding animals ....(Illustrated) 1.50

Hints to Horsekeepers. By H. W. HERBERT. How to breed, buy, break, drive, ride, groom, use, feed and physic ....(Illustrated) 1.75

The Family Horse. By G. A. MARTIN. Stabling, care, feeding, working, driving, etc. (Nothing about breeding.) ....(Illustrated) 1.00

The Saddle Horse. Riding, training and feats under saddle ....(Illustrated) 1.00

Handbook of the Turf. By S. L. BOARDMAN. Trotting rules and other information. 1.00

Practical Horseshoer. Shapes for different feet; interfering, overreaching, contraction, diseases, tools, methods for handling the vicious, etc ....(Illustrated) 1.00

How to Handle and Educate Vicious Horses. By O. R. GLEASON..(Illustrated) .50

Training Trotting Horses. By CHAS. MARVIN. For colts and horses.(Illustrated) 3.50

Scientific Horseshoeing. By PROF. WM. RUSSELL. Enlarged edition of 1895. Leveling and balancing action; curing diseases, etc ....(Illustrated) 4.00

Modern Horse Doctor. By G. H. DADD, M.D., V. S. Preservation and restoration of health; treatment of lameness, etc ....(Illustrated) 1.50

Riley on the Mule. By HARVEY RILEY. Feeding, training, uses, etc..(Illustrated) 1.50

CATTLE.

Allen's American Cattle. By LEWIS F. ALLEN. A standard authority. History of breeds, breeding, management and improvement ....(Illustrated) $2.50

Cattle Breeding. By WM. WARFIELD. By common consent the most valuable and practical American work on the subject ....(Illustrated) 2.00

Cattle; their Management in Health and Disease. By GEO. ARMATAGE, M.R.C.V.S. (London). A guide for the farmer and breeder; diseases and treatment 1.00

Manual of Cattle Feeding. By H. P. ARMSBY, Chemist, Connecticut Agricultural Experiment Station. Laws of nutrition; feeding stuffs and feeding...(Illustrated) 1.75

SHEEP AND SWINE.

Shepherd's Manual. By HENRY STEWART. A valuable treatise on sheep for American farmers; breeds, breeding, management and diseases ....(Illustrated) $1.50

The American Merino for Wool or for Mutton. By STEPHEN POWERS. Selection, care, breeding and diseases; a valuable treatise ....(Illustrated) 1.50

Sheep; their Management in Health and Disease. By GEO. ARMATAGE, M.R.C.V.S. (London). A guide for the sheep farmer; maladies, causes, remedies 1.00

Swine Husbandry. By F. D. COBURN. Revised and enlarged edition. Breeding, rearing, management, diseases, prevention, treatment ....(Illustrated) 1.75

Harris on the Pig. By JOS. HARRIS. Various breeds discussed; management, etc. 1.50

POULTRY.

Profits in Poultry and Profitable Management. The experience of practical men in all departments; useful and ornamental breeds ....(Illustrated) $1.00

Practical Poultry Keeper. By L. WRIGHT. A complete and standard guide for domestic use, market and exhibition ....(Illustrated) 2.00

The American Standard of Perfection. (Adopted by American Poultry Assn.) Descriptions of recognized breeds, judges' instructions, etc ....(Illustrated) 1.00

An Egg Farm. By H. H. STODDARD. Management of poultry in large numbers .... .50

Five Hundred Questions and Answers in Poultry Raising. Also feed, care, diseases, eggs, incubation, buildings, etc .... .25

Capons for Profit. By T. GREINER. How to make and manage; plain instructions for beginners ....(Illustrated) .30

Turkeys and How to Grow Them. By HERBERT MYRICK, and Essays from Practical Growers. History, breeds, successful management, etc ....(Illustrated) 1.00

Duck Culture. By JAS. RANKIN. Natural and artificial ....(Illustrated) .50

Low-Cost Poultry Houses. By J. W. DARROW. Plans and specifications for $25 to $100 buildings; other conveniences .... .25

Incubators and their Management. By J. H. SUTCLIFF ....(Illustrated) .40

APIARY, BEES, HONEY.

A B C of Bee Culture. By A. I. ROOT. A cyclopædia on bees, honey, hives, implements; honey plants, etc ....(Illustrated) $1.25

Quinby's New Bee-keeping. By L. C. ROOT. The mysteries explained; 50 years' experience; latest discoveries and inventions 1.50

Bees and Bee-Keeping, Scientific and Practical. By F. R. CHESHIRE. An exhaustive treatise on advanced bee culture. 2 volumes 6.40

DOGS, CATS, RABBITS.

Dogs of Great Britain, America and Other Countries. Breeding, training, management, diseases, noted dogs, best hunting grounds, etc ....(Illustrated) $2.00

Cats, Domestic and Fancy. By J. JENNINGS. Varieties, breeding, management, diseases, remedies, exhibiting, judging ....(Illustrated) 1.00

Practical Rabbit-Keeper. By CUPICULUS. Species, raising for pleasure or profit; courts, warrens, hutches, fencing, etc ....(Illustrated) 1.50

BIRDS.

American Bird Fancier. By MESSRS. BROWNE and WALKER. A complete manual on breeding and rearing song and domestic birds ....(Illustrated) $0.50

Our Common Birds and How to Know Them. By J. B. GRANT. Enables any one to recognize and name them ....(Illustrated) 1.50

Canary Birds. A manual of information .... .50

Canaries and Cage Birds. By GEO. H. HOLDEN. Food, care, breeding, diseases and treatment of all house birds for pleasure or profit ....(Illustrated) 2.00

Diseases of Cage Birds. By W. T. GREEN. Causes, symptoms, treatment .... .40

Pigeon-Keeping for Amateurs. By J. C. LYELL. A complete guide ....(Illustrated) 1.00

Game Birds of North America; How to Know Them. By BATES. (Illustrated) 1.25

AQUARIA, FISH AND FISH RAISING.

Amateur Aquarist. By M. SAMUEL. Fresh water aquariums ....(Illustrated) $1.00

The Goldfish and Its Culture. By H. MULERTT. Breeding, raising, enemies, diseases, ponds, etc ....(Illustrated) 1.00

American Fish Culture. By T. NORRIS. Details of, breeding, etc ....(Illustrated) 1.75

Home Fishing and Home Waters. By SETH GREEN. Utilization of farm streams; artificial ponds; transportation of eggs and fry, etc .... .50

SPORT—FISHING, HUNTING, TRAPPING, SAILING.

The Scientific Angler. By DAVID FOSTER. Artistic angling $1.50

American Fishes. By G. B. GOODE. Game and food fishes; habits and methods of capture ....(Illustrated) 3.50

American Game Bird Shooting. By J. M. MURPHY. Haunts, habits, bagging, luring, devices; dogs, guns, etc ....(Illustrated) 2.00

Hunter and Trapper. By H. THRASHER. Best methods for foxes, deer, bears, etc. .75

Camp Life in the Woods. By W. H. GIBSON. Tricks of trapping and trap making, bait receipts, camp shelter, boat making; tanning fur skins ....(Illustrated) 1.00

Taxidermy. By W. T. HORNADAY. Hand-book for the amateur; preserving and setting up animals, insects, etc ....(Illustrated) 2.50

American Boys' Book of Sport. By D. C. BEARD. Outdoor games for all seasons; over 300 illustrations ....(Illustrated) 2.50

American Boys' Handy Book. By D. C. BEARD. What to do and how to do it. (Liberally illustrated) 2.00

Boat Building for Amateurs. By MESSRS. KEMP and NEISON. Instructions, diagrams; skiffs, canoes, sailboats, handling, etc ....(Illustrated) 1.00

Practical Boat Sailing. By D. FRAZER. Management of small boats and yachts, emergencies, nautical terms, etc ....(Illustrated) 1.00

HOUSEHOLD BOOKS.

Canning and Preserving. By MRS. RORER. How to can fruits and vegetables, make preserves, marmalades, fruit butter; dry fruits and herbs, etc $0.40

Fruits and How to Use Them. By MRS. POOLE. Nearly 700 receipts for preparing various fruits in various forms and how to use them ....(Illustrated) 1.00

Canning and Preserving Fruits and Vegetables. By ERMENTINE YOUNG. Also fruit pastes, syrups, evaporating fruits, etc .... .25

The National Cook Book. By MARION HARLAND and C. T. HERRICK. An entirely new work 1.50

The Dinner Year Book 1.75

American Dainties and How to Prepare Them. By an American lady .... .40

Common Sense in the Household. New and revised edition. A manual of practical housewifery 1.50

Everybody's Paint Book. All about polishing, painting, staining, kalsomining, renovating furniture, etc ....(Illustrated) 1.00

Toy Making for Amateurs. By JAS. LUKIN. Home construction of wooden toys; simple and driven by various contrivances ....(Illustrated) 1.00

American Girl's Handy Book. By THE MISSES BEARD. How to amuse yourself and others ....(300 illustrations) 2.00

The Language of Flowers. By J. INGRAHAM. Includes floral poetry. Small ed., 50c.; cloth, $1.00 ....gilt, 1.50

The Art of Skeletonizing Leaves and Seed Vessels. Skeletonizing, bleaching, coloring and forming "phantom bouquets." ....(Illustrated) 1.00

Cane Basket Work. By ANNIE FRITH. A practical manual on weaving fancy and useful baskets ....(Illustrated) 1.00

We will procure for our customers any book published, if still in print, and deliver free in the U. S. at the regular retail price.

THE "HENDERSON" Chilled Plow

A Grand Plow.
Lightest Draft.
Runs Steady.
Correct Shape.
Harder than Steel.

THE OUTLINES

and shape of the mouldboard and castings are exactly right to do perfect work with the least possible draft, and the greatest ease to the plowman.

THE CHILLED MOULDBOARDS

are harder than the hardest tempered steel. Rust will not eat into or roughen them, and the effects of exposure to the weather for weeks will be obliterated by turning a few feet of soil, leaving the surface like polished glass; they will scour in any soil from sharp gravel to muck.

THE CUTTING EDGE, OR "SHIN PIECE,"

is separate from the mouldboard (*on all sizes larger than Medium One Horse*), and can be taken off and ground sharp or be replaced by a new one at small cost, making an old plow as good as new. Farmers who understand that half of the draft is expended in cutting the furrow slice will realize this great advantage over other plows, the mouldboards of which, being made in one piece, cannot be remedied when the cutting edge begins to wear blunt and round, causing them to draw hard and do poor work.

THE LANDSIDE

inclines in at the bottom away from the unplowed ground, thereby avoiding unnecessary pressure and friction, reducing the draft, and, besides, cutting a furrow of such a shape that—instead of having to be forced over on its face by the heel of the mouldboard, as is the case with ordinary plows—it falls by its own weight as soon as it passes a perpendicular position, thereby relieving the plow of that much labor.

THE HEEL OF THE BEAM

is adjustable on a slotted brace between the handles, so that the plow can be regulated to make a lapped, flat, small, or large furrow by simply loosening *one* nut and shifting the beam, which, being placed over the centre of the draft of the mouldboard, and not stationary over the "landside," as is customary, relieves the plow of all unnecessary bottom and side friction and permits such perfect adjustment that it can be balanced to run steadily and without that physical force required by the plowman to hold rigid beam plows to their work. It is also a desirable feature in one-horse plows, as it allows the horse to walk in the furrow when desired.

THE STANDARD

is placed inside of the cutting line, and is of such a shape that it does not choke.

THE WHEEL

can be swiveled around to run in a direct line with the plow, no matter which way the beam is shifted; this makes the plow draw always perfectly true.

THE SKIM, OR JOINTER,

fairly pays for itself on every acre of sod-ground plowed. It turns under all weeds, manure, stubble and surface soil, making a miniature furrow over which the main furrow is thrown, so that it breaks over, cracking, opening, and thereby mellowing the soil by its better exposure to the air. The straw, manure, etc., being entirely covered, decomposes more rapidly; and the edges, not being exposed, do not interfere with the successful working of the harrow.

"THE PATENT REVERSIBLE SELF-SHARPENING SLIP SHARE"

we recommend to be used on all ordinary soils (very rocky or sticky soils excepted). All farmers know that the point wears round, or "sledge-runner like," long before the wing of the share is worn. By using the PATENT SLIP SHARE the point can not only be turned over when it wears dull, but when worn out a new one can be inserted in a minute's time. One share will outwear several points and one share with three extra slip-points will do the work of EIGHT solid shares, and do it better and with more ease, for the plow will always run level and the points keep sharp all the time, by simply turning over the slip when necessary until worn out.

DEEP SUCTION SHARES

can be furnished for the Two and Three-horse plows, to be used in extreme cases where the ground is particularly hard and when ordinary shares do not have "suction" or "bite" enough to keep the plow down to its work.

THE "HENDERSON" CHILLED PLOW

is better made, more durable, and more highly finished than any plow in the market. Nothing but the very best material is used in its manufacture. Every plow gets three coats of paint and varnish, and is warranted to give satisfaction.

SLIP SHARE.

NUMBER.	SIZES OF THE "HENDERSON" CHILLED PLOW.	Price, Plain.	With Wheel.	With Wheel and Jointer or Skim.	Solid Share.	Slip Share with Slip.	Extra Slips.	Letter on Slip.
A-3	Light, one horse, - - - Cuts a furrow 4½ by 9 inches.	$4.50	....	....	$0.20	$0.25	$0.08	X
B-3	Medium, one horse, - - " " 5 by 10 "	5.00	$6.00	....	.20	.25	.08	X
13	Full, one horse, - - - " " 5½ by 11 "	6.50	7.50	....	.25	.35	.08	A
D-3	Light, two horse, - - - " " 6 by 12 "	7.50	8.50	$10.50	.30	.40	.10	C
23	Medium, two horse, - - " " 7 by 13 "	8.00	9.00	11.00	.35	.45	.10	D
E-4	Full, two horse, - - - - " " 8 by 14 "	8.50	9.50	11.50	.35	.45	.10	E
43-A	Heavy, two horse (or three horse) " " 9 by 15 "	9.00	10.00	12.00	.35	.45	.10	E

An extra Share goes with each plow with solid shares. **An extra "Slip"** with each plow with SLIP shares. **A Wrench** free with each plow.

THE HENDERSON STEEL PLOW.
STEEL BEAM AND STEEL MOULD.

Made of genuine fine tempered steel. Scours in stickiest soil, never breaks, and is one-quarter lighter for man and team. Similar in shape to the celebrated Scotch plows, with short beam and long, low adjustable handles. The "mould" and "land" are perfection in shape. Contains all improvements, separate shin piece, new clevis, quickly adjusted to run the plow shallow or deep, wide or narrow. Warranted to give highest satisfaction.

Price, two-horse, furrow, 5 to 7 inches deep by 12 to 15 inches wide, steel beam, steel mould and chilled share, and extra share.............$12.00
With wheel and jointer.. 14.00
With steel share in place of the two carbonate shares furnished with
   the above, $1.50 extra.

THE "BOSS" ONE-HORSE PLOW.

THE PLOW FOR GARDENERS AND TRUCKERS.

The best One-Horse Turn Plow ever sold. It makes a beautiful furrow, turning everything under—weeds, grass, trash and all. We call attention to the fact that the purchaser actually gets four plows for the price of one, viz.: One complete plow, No. 2½ mouldboard and share, making a large one-horse plow. A No. 1½ mouldboard and share, making a medium-size one-horse plow. A cabbage mouldboard and share, making a complete cabbage plow, and by using the plow without mouldboard, with the small share bolted to standard, makes a superior plow for working among cabbage, strawberries, etc. Price, $6.00, or with a splendid *double* mouldboard for furrowing, hilling, etc., price, $7.50.

CHEAP "CAST-IRON" PLOWS.

Good general purpose plows. Wood beams, ground castings.

Light one-horse,	No. 18	.........price,	$3.25
Medium one-horse,	" 18½	.... "	3.50
One-horse, large,	" 19	.... "	4.00
Light two-horse,	" 19½	.... "	4.50
Two-horse,	" 20	.... "	5.00
Heavy two-horse,	" 21	.... "	6.00
Side Coulter and Clamp, extra................			1.25
Wheel and Attachment, "		....	1.25

CORN PLOW.

Largely used in some sections for marking out corn and potato ground, and also for plowing.

Cast Iron, 1 handle, weighs 50 lbs...........$5.00
  "   2 handles........................... 5.50
Steel, 1 handle, weighs 40 lbs.......... 5.75
  "   2 handles.......................... 6.25

THE NEW "IMPERIAL" SULKY PLOW.
FOR TWO HORSES.

The Imperial Sulky Plow is a combination of great strength and simplicity, expressly built for hard service. By means of the levers the operator has complete control of the plow at all times. The Spring Hoist is of great assistance in handling the plow. The rear wheel is held in position by a lock, which is released by slight pressure of the foot for turning at the corner.

The steel beam is long and high in the throat, which prevents clogging. The hitch is directly to end of beam. It turns a square corner, right or left, on the point of the share, the plow not being hoisted from the ground while turning.

We guarantee it to do more and better work than any sulky plow ever offered. Weight (as shown in cut), 475 lbs.

It is not intended to be used with pole and yoke, although we furnish same on order at an additional price, when required.

No.	Cut.	Mould and Landside.	Share.	Price.
D 12.	14-inch.	Chilled.	Cast.	$34.00
DC 12.	14- "	Steel.	Steel.	38.00

Jointer, $2.00 extra.   Rolling Cutter, $3.00 extra.
Fin Cutter, 75c. extra.

EXTRA EQUIPMENTS (specify in order if wanted).
Double-trees, $2.00 extra.   Triple-trees, $3.50 extra.
Pole and Yoke, $3.00 extra.

"CABBAGE" OR "SKELETON" PLOW.

This is used principally for cultivating cabbage, corn and other leafy crops. The wing of the mouldboard being omitted, allows the plow to run close to the plant, under the spreading leaves, without injuring or throwing earth over them.

Price, medium size, No. 1½ $3.50
   full size,   " 2   .... 4.00

DOUBLE MOULD PLOW.

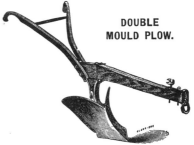

For opening furrows or making drills to plant potatoes, corn, etc.; for ditching, hilling, ridging, breaking out middles, etc.

Light	one-horse, iron,	$4.00;	steel,	$5.00
Medium	"	"	4.50;	" 6.00
Heavy	"	"	5.00;	" 7.00
Light	two-horse, "		5.50;	" 7.75
Medium	"	"	6.00;	" 8.25
Heavy	"	"	7.50;	" 10.00
Wheels, extra, for one-horse plows.......				.75
" " " two-horse "			....	1.00

STEEL.

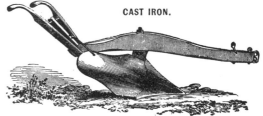

REVERSIBLE PLOWS,
FOR LEVEL LAND OR SIDE HILL.

The mouldboard is made to be turned to the right or left, alternately, for the purpose of throwing the furrow-slice down hill, whichever way the team may be passing. It is also used for level land, to avoid dead furrows and ridges without plowing around the field.

CAST IRON.

STEEL REVERSIBLE (Level Land or Side Hill) PLOWS.

This plow has a steel beam, a long steel mouldboard, will turn a furrow on level land equal to any plow in use, and does grand work on hillsides. Light of draft and light to handle. It has a side-shifting clevis, is operated between handles, and the taking of land by plow can be regulated while in operation. The jointer turns automatically, with the mould always in direct line of cut. The foot latch is automatic and positive in action and locks firmly, overcoming any possibility of unlatching when plowing on stony soil.
No. X, two-horse, cuts 12 to 14 inches. Plain, $13.00; with wheel, $14.00; with wheel and jointer, $16.00.
Automatic shifting coulter supplied in place of jointer if preferred.

CAST (Side Hill) PLOWS.

No.		Plain.	For Wheel add	For Cutter add	For Draft Rod, add
0	Light Horse.........	$4.25	$1.00	....	....
00	One- "	5.25	1.00	....	....
B 1	Two- "	7.00	1.00	$1.00	$1.25
A 1½	Sod.........................	7.75	1.00	1.00	1.25
A 2	"	8.00	1.00	1.00	1.25
A 3	" large.................	8.50	1.00	1.00	1.25

THE IMPROVED WINGED SHOVEL PLOW

For furrowing out, hilling and cultivating all kinds of crops that are planted in rows. Cuts up all grass and weeds. Wings and points of steel. The wings can be opened or closed to throw off more or less earth, and to adapt them to rows of different widths. For weeding, and when it is not desired to throw soil against the plants, the wings should be reversed. The soil then passes over the plow. The depth is regulated by the wheel and draft hook. Handles adjustable in height. Price, $7.00, or with the

Potato Digger Attachment, $10.50. By the changing of one bolt it may be converted into a first-class potato digger. The fingers are adjustable and removable.

MINER'S GOLD MEDAL SUB-SOIL PLOW

By following the plow with a sub-soil plow the earth can be broken to a depth of from 15 to 20 inches, giving roots a wider range for food, and the plants are hardly affected by excessive drought or a wet season, as the deep soil absorbs all the rain like a sponge and gradually gives off moisture during dry, hot weather.

The "Gold Medal" involves new principles and accomplishes the work without throwing any sub-soil on top. Its merits are ease of penetration, light draft, superior pulverization of hard pan, perfection of the "mole track" effect. The one-horse plow will reach to a depth equal to the height of its standard—15 to 16 inches—and the two-horse plow to the depth of 20 inches. Prices for one horse.................................$6.00; with wheel, $7.00
" " two horses................................ 8.00; " 9.00

NATIONAL REVERSIBLE SULKY PLOW.

Of light draught and easily handled. The only successful sulky plow made that will plow stony, rough, side hill or level land; especially designed to meet the wants of Eastern farmers, and succeeds not only upon level farms, leaving them without tracks or dead furrows, but is equally adapted to stony, rough, side hill farms, turning the land with the slope, and not up the hill, as is the case with all one-way sulky plows. We believe this to be the only sulky plow in the world that can justly lay claim to being a success for all kinds of work and in all varieties of soil. As shown in the cut, a right and left steel flat-land plow is mounted upon a steel beam, one being at right angles with the other, and easily revolved by unlocking a hand lever at the rear of the driver, the weight of the upper plow causing the lower to raise. Each plow has an easy adjustment to make it cut a wide or narrow furrow, and is raised out of the ground by a perfect working power lift, and set in again by a foot lever, so that the operator has both hands with which to manage his team. The adjustable seat enables the operator to always sit in a level position and on the uppermost side in plowing side hill land. Price, including extra points, neck yoke, evener, whiffletrees and jointer, $55.00. *If preferred, we will send in place of jointer either straight or rolling coulters.* Either plow can be detached from the sulky and fitted with beam and handles, thus making two perfect flat-land plows. For this purpose we furnish beam, wheel, clevis and handles for $5.00.

SINGLE SHOVEL PLOW.

For cultivating corn, potatoes, etc., hilling, furrowing, marking out, etc.; wrought-iron beam. Price, $3.00.

DOUBLE SHOVEL PLOW.

An effectual pulverizer; destroys weeds, and is unexcelled for corn cultivation; it will not clog or choke; it runs easy and steady; weighs only 38 to 40 lbs.; wrought-iron beams. Price, $3.50.

ALL STEEL "GRUB OR ROOT GROUND" PLOW.

For use in rough, rooty or new ground. Made with steel beam, mold and coulter. The illustration shows curved coulter; they are now made straight. Handles adjustable in height. A strong and serviceable instrument. Price, $10.00.

WROUGHT IRON GRUB HOOK OR STONE AND ROOT PULLER.

Land can be cleared of impediments to thorough cultivation by this grub hook. Its peculiar shape causes it to operate as a lever when the team is hitched close to the beam, and it has a great power in rolling out stones and tearing out roots, stumps and logs. One man with team can accomplish as much in one day with this grub hook as four men with crowbars, levers, etc., in a week. Price, $6.00.

"PLANET, JR." IRISH POTATO DIGGERS.

The "Planet, Jr." Double Moldboard Potato Digger is simple, easily understood and worked; has no gearing or shakers; the draught is light and there is nothing to wear out but the shares, and they and the moldboards are hardened steel. The action is that of raising, dividing and turning the row on edge, and then separating the potatoes from the earth by means of steel fingers and lifting them to the surface. The vine turner makes it perfectly easy to dig when the vines are green, and is of great assistance in weedy patches. Price, $16.00.

The "Planet, Jr." Single Moldboard Potato Digger is constructed exactly like the above, except it throws but one way. The double leaves the ground more level, but the single has the advantage of leaving the potatoes less scattered. Price, $16.00.

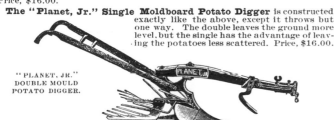

"PLANET, JR." DOUBLE MOULD POTATO DIGGER.

THE "PLANET, JR." SWEET POTATO DIGGER.

The recognized standard for sweet potato digging; long moldboard with steel digging tines; beam is high, giving ample room for the gang of vine cutters, so that we use a high arch. The steel discs can be set 16 or 18 inches apart and at any depth or angle desired, and each can be adjusted for depth independently of the other. The moldboards are adjustable for digging in different soils and all width rows, and in both hills and drills. At a single passage the tool cuts the vines on both sides of the row, plows out the potatoes and leaves them on top of the furrow slice, most of the earth being loosened or knocked off, and the crop left in the best condition for gathering into heaps. Price, $16.00.

SCOTCH HINGE HARROW.

THE OLD RELIABLE FIELD HARROW, GOOD ON ANY KIND OF LAND.

It is so coupled that each section has a vibratory motion independent of the other, and consequently adjusts itself to the uneven ground, and either side may be lifted while in motion to free it from weeds, stones, etc. The draft is adjusted so no two teeth follow in the same track. It is simple, durable, and well finished.

Prices, 40 teeth, $14.00; 50 teeth, $16.00.

REVERSIBLE WOOD FRAME SMOOTHING HARROW.

The teeth of round steel are so attached to frame that by hitching to one end of the harrow they pull straight, operating like an ordinary spike-tooth harrow; by hitching to the other end the teeth pull slanting, making one of the best smoothing harrows on the market. In this form it levels, pulverizes and mellows the soil without turning up weeds, trash, etc., and will not clog; and it acts well for covering grain or grass seeds; in this form it is also valuable for cultivating, without injury, while small, grain, corn, etc.

PRICES.

Size.		Size.	
One-horse, 1 section, 24 teeth,	$8.50	Two-horse, 3 sections, 72 teeth,	$18.00
Two- " 2 sections, 48 "	12.00	Three- " 4 " 96 "	25.00

COMMON SQUARE HARROW.

The teeth are tapering in shape, so that when loosened by weather or rough usage they can be driven tight again. The teeth bars, being riveted at their ends, cannot split.

PRICES.

One-horse size, made light, has 25 teeth, $7.50
Two- " " " larger and heavier.
  36 teeth 12.00

EVANS' GARDENER'S HARROW.

It is designed for one horse, convenient to handle and get around garden plots, etc. Beams of oak, teeth of steel. For transportation to and from the field invert the harrow on the runners, which saves lifting and loading.

PRICES.

24 teeth $8.00
30 " 9.00

The "Acme" Pulverizing Harrow, Clod Crusher and Leveller.

No. 17.
"ACME" HARROW.

The best general-purpose size.

A general-purpose harrow, that will crush, cut, lift, turn, smooth and level the soil to perfection, all in one operation. While preëminently adapted for heavy, stubborn land, it can be adjusted by means of the levers and runners to do perfect work on the lightest soil. It cuts the entire surface without disturbing sod or trash that has been turned under by the plow. It prepares a perfect seed bed, and also covers seed in the best manner. The draft is reduced to a minimum. Being made entirely of cast steel and wrought iron, it is practically indestructible. Nothing but the coulters can possibly wear, and these are readily replaced at a trifling cost.

PRICES AND SIZES.

No. **17.** Riding Harrow. Two-horse size. 6½-foot cut with 2 runners for transporting............................$15.00
No. **20.** Walking Harrow. (No seat, elevating lever or spurs.) Two horses. Cuts 6½ feet................... 10.00
No. **18.** Riding Harrow. Three or four horses. Cuts 13½ feet 30.00
Two-horse whiffletree and neck yoke.................Extra, 4.00
Three-horse drawbar......$1.50. One-horse whiffletree, .50

BUTTERFLY HARROW.

Frame joined by hinges, so it adapts itself to an uneven surface, and either side may be elevated to free it from stones, sods, etc.; or may be folded upon the other in passing between stumps, trees, etc. The teeth draw at equal distances apart.

PRICES.

14 teeth $9.00	26 teeth........$16.00
18 " 10.00	30 " 18.00
22 " 14.00	

"ACME" One=horse Harrow.

No. **H.** Works 4⅓ feet wide. 8 coulters. Price, with seat and lever adjusted, $13.00
No. **G.** A splendid one-horse cultivator. 6 coulters. Cuts 3 feet wide. Has handles, as shown in cut, but no seat or lever..Price, $8.00

ALL STEEL SPRING = TOOTH FLOAT HARROW.

Spring-tooth harrows and cultivators are in great favor with those who have tried them. They are strong, light of draft, and aided by the vibratory motion of the teeth they pulverize and level the soil better than almost any other style. This Float harrow, while the cheapest style of spring-tooth harrow yet, is thoroughly satisfactory. The teeth of tempered steel are adjustable.
Prices, 14-tooth, $10.00; 16-tooth, $11.00; 18-tooth, $12.00.

EMPIRE SPRING=TOOTH RIDING=WHEEL HARROW.

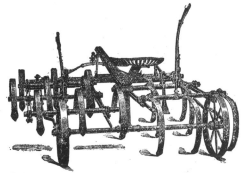

This is the highest development in spring-tooth harrows. The driver operates the levers from the seat, changing angle and depth at pleasure, or throws the teeth entirely out of the soil. The weight of the frame and driver, being carried on the wheels, lessens draft. Each half of the harrow is operated independently of the other from the seat.
Prices, 16-tooth, $22.00; 18-tooth, $24.00; 20-tooth, $26.00.

NEW LEVER=SET SPRING HARROW.

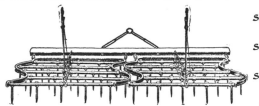

A
SMOOTHING
HARROW
A
SPIKE-TOOTH
HARROW
A
SPRING=TOOTH
HARROW

ALL IN ONE.

The only Harrow in the world having Spring Action upon Spike Teeth.

We accomplish this by a heavy Coiled Spring on the bars connected with the Lever Adjuster. This spring allows the teeth to yield in meeting an obstruction, thus saving strain or breakage, as well as allowing a constant vibration to the teeth.

We thus combine all the advantages of a Spring Tooth and the finer cutting of Spike Teeth in one Harrow, an advantage readily appreciated by users of Harrows. It can also be set, teeth slanting backward, as a smoothing Harrow. Made entirely of steel, handsome, strong, simple, and does grand work.

Prices : With 25 Teeth in One Section, spreading 4 ft., $9.00
" 50 " Two " " 8 " 15.00
" 75 " Three " " 12 " 22.50

THE MEEKER DISC....

....SMOOTHING HARROW.

It is a splendid substitute for a rake for garden purposes, or for seeding down. It pulverizes and grades, leaving the ground in as fine a condition as could possibly be done by hand. No market gardener or farmer can afford to be without it.

The frame is 6⅔ ft. x 6 ft., with four sets of rollers having 58 discs. The discs on each set of rollers work between each other.

The board in the centre is set at an angle; is adjusted up and down, and acts as a leveller.

The discs grind all lumps so fine that seed must come up, and mashes small stones below the surface better than any field roller; it levels the ground at the same time, which a field roller does not do.

Price : $20.00

....COMBINED....

FORCE FEED SEEDER and **DISC HARROW.**

Sows all kinds of grains in any quantity desired. A perfect force feed. Thus you harrow, sow and cover all at one operation. The Seeder is easily attached or detached.

Price : 12 discs, Harrow and Seeder, $60.00. Cuts 6 ft. wide.

Cut=a=way or Spading Harrow.

This style of Disc Harrow has many friends. Sections of the discs being cut away gives a longer cutting edge and cuts and buries weeds and trash deeply, and pulverizes the soil to perfection; the cutting angle is regulated by a lever.

Price : 2-horse size. 12-16 in. discs, cuts 6 ft. wide, $25.00.

NEW FLEXIBLE DISC HARROW.

The revolving sharp-edged, circular, concave discs set with the lever at any desired angle in relation to the line of drafts ; they fully pulverize the soil by cutting, lifting and turning it over in fine, small furrows. It is the best harrow made for sward-land, as it will not turn up the sod—catch roots, stubble, manure, nor clog—but will cut and mix them with the soil. It will work on wet and adhesive soil where no other implement will. It is a good clod crusher, and it excels in putting in grain, leaving it in little drills at the proper depth and covering it well.

The New Flexible Disc Harrow is the only Disc Harrow having *independent adjustable spring pressure* upon the inner end of each gang of discs, allowing any amount of pressure to be thrown upon the inner ends of the gangs by the foot of the operator.

The scrapers are held at any point, by means of spring pressure upon the sliding bars.

2-Horse. No. 3. Has 12-16 in. Discs, cuts 6 ft...$25.00
1-Horse. No. 0. Has 6-16 in. Discs, cuts 3 ft...... 21.00

REVOLVING HARROW.

This revolves continually while in action, so the teeth cut and pulverize every inch of soil ; besides, it cannot clog and is, in consequence, invaluable in ground littered with corn butts, stones, turf, roots, etc. It is a splendid leveller and grain coverer. The heavy iron rudder wheel keeps the harrow immediately behind the team.

Prices : 1-horse, 20 teeth, $20.00
2- " 30 ". 25.00

The Planet, Jr., All Steel Leveller.

This is a perfect tool for fining, levelling and preparing seed beds and planting ground. Its width is 4 ft. 6 in.; its weight, 100 lbs. It can be instantly regulated to cut deep or shallow, and the cutting bars and crushing and smoothing plates make it efficient in surface work. It is invaluable for market gardeners.

Price : $12.00.

SPRING=TOOTH CULTIVATOR.

Spring - Tooth Cultivators and Harrows are becoming very popular. There is just enough vibration to the teeth to thoroughly pulverize the soil, and they tend to leave the weeds on top of the soil. They are especially useful in truck gardens and vineyards and on stony ground. Frame of wrought iron, and is adjustable in width; teeth of oil-tempered steel, are interchangeable and adjustable for angle and depth.

Prices
(complete with wheel) :
Pony size, 6 teeth, $5.50
1-horse size, 8 teeth, $6.75

DIAMOND=TOOTH "HARROW=CULTIVATOR."

A splendid tool for cultivating garden and field crops in rows. The numerous diamond-shaped steel teeth thoroughly pulverize and loosen the soil without throwing earth on the plants. The frame expands or contracts by operating the lever between the handles, even while the horse is in motion, adapting the Cultivator to rows of varying widths. The back, wide, flat, steel hoe sweep cuts off all weeds. Can be used or not as the work requires.

Prices : Plain, $4.50 ; with Wheel, $5.00 ; Sweep extra, 75c.

THE PLANET, Jr., PIVOT WHEEL RIDING CULTIVATOR, PLOW, MARKER —and RIDGER.

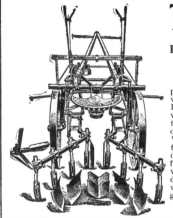

Will do almost every kind of work possible in the cultivation of crops, with the greatest ease and perfection. It is comfortable to ride upon, light in weight, light in draft, simple in operation and durable in every part; it will cultivate, furrow out, hill up, etc. Wheels close to 30 inches, and widen to 40 inches, adapting it to rows of different widths; hoes will slide and fasten in any position, does accurate work without constant watching; depth regulation is by lever combined with spring tension; improved plant guards.

Price, - - $40.00.

PLANET, Jr., UNIVERSAL CULTIVATOR, HARROW, MARKER and COVERER.

Gives excellent satisfaction in a great variety of work; in fact, it is a **universal tool holder;** almost every shape of tooth we make fits it, so that this implement will perform an immense variety of work at all seasons, and in nearly all crops. (*Complete description mailed on application.*)

PRICES.

As shown, cutting 5 ft. 3 in.,	$35.00	Furrowing Shovels, set of 3, -	$1.00
One Horse, cutting 4 ft., - -	30.00	Turning Plows, set of 4, - -	3.00
Extensions, extra, add'g 15 in.,	4.00	Leveller, 5 ft. 6 in., - - - -	4.00
Set of nine 10 in. Sweeps, - -	3.15	" 7 ft., - - - - -	5.00

PLANET, Jr., "DEEP TILLAGE" CULTIVATOR.

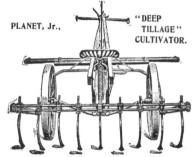

Will cultivate up to 10 inches in depth and leave the ground perfectly level.

Price, - - $31.00.

"SPRING-TOOTH ORCHARD" CULTIVATOR.

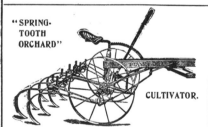

Spring teeth are looked upon with favor for use in orchards and stumpy or rough ground. We have aimed to produce a shape that will allow the two rows of teeth to be widely separated *so as not to choke,* and at the same time have all *attached to the same bar;* and also to provide an easy method of having the teeth work at the same depth and angle, and therefore with **the same efficiency, as they wear shorter.**

Price, - - $31.00.

THE "DIAMOND" SULKY or RIDING CULTIVATOR.

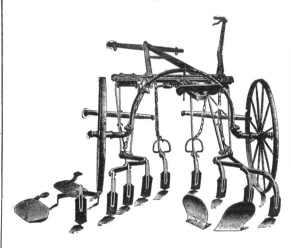

THIS Cultivator is adapted for light, sandy soil, and is especially suited for truck garden cultivation. The centre drag bars have free and independent action up and down or sideways, allowing the operator to work them near to or far from the row, at will, and with perfect control. Adjustable teeth, independent drag bars, flexible hitch, quick levelling device, solid arch, adjustable track. Width of track, 3 ft. 4 in., adjustable to 4 ft. Furnished complete, with eight reversible shovels, hilling moldboards, and centre attachments.

Price, - - - $24.00.

PLANET, Jr., 4-ROW BEET HOE.

Hoes four rows at once, 14, 16, 18 or 20 inches apart, or three rows 22, 24 or 26 inches apart. The wheels adjust to suit; all improvements. **Price, $50.00.**

STEEL "DISC" CULTIVATOR.

Adjusts to any angle; sets by lever to throw soil toward or from the row; sets to cut deep in centre and shallow next the row, or reverse, or level; adjusts to any width; rowed crops can be hilled thoroughly; effectually cuts up and buries weeds.

Price, with three 16-inch discs on each side, **$26.00.**

The STANDARD SPRING-TOOTH WHEEL CULTIVATOR,
Furnished either as a Walking or Riding Cultivator.

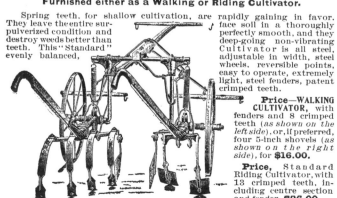

Spring teeth, for shallow cultivation, are rapidly gaining in favor. They leave the entire surface soil in a thoroughly pulverized condition and perfectly smooth, and they destroy weeds better than deep-going non-vibrating teeth. This "Standard" Cultivator is all steel, evenly balanced, adjustable in width, steel wheels, reversible points, easy to operate, extremely light, steel fenders, patent crimped teeth.

Price—WALKING CULTIVATOR, with fenders and 8 crimped teeth (*as shown on the left side*), or, if preferred, four 5-inch shovels (*as shown on the right side*), for **$16.00.**

Price, Standard Riding Cultivator, with 13 crimped teeth, including centre section and fender, **$26.00.**

THE "PENN" SULKY CULTIVATOR
And DOUBLE ROW CORN PLANTER and FERTILIZER SOWER.

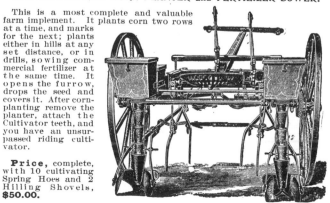

This is a most complete and valuable farm implement. It plants corn two rows at a time, and marks for the next; plants either in hills at any set distance, or in drills, sowing commercial fertilizer at the same time. It opens the furrow, drops the seed and covers it. After corn-planting remove the planter, attach the Cultivator teeth, and you have an unsurpassed riding cultivator.

Price, complete, with 10 cultivating Spring Hoes and 2 Hilling Shovels, **$50.00.**

HENDERSON'S LOW-PRICE "Leader" Cultivator.

HENDERSON'S Low-Price "LEADER" CULTIVATOR.

Wrought-Iron Expanding Frame.

This is a very popular cultivator, a first-class implement, strong, well finished and light; it can be contracted or expanded to suit any width of row; the standards can be adjusted at different angles, and the steel teeth when worn can be reversed.

Price, plain, as shown in the cut, $3.25 ; or with wheel, $3.75.

HENDERSON'S LOW-PRICE LEADER HORSE HOE.

Wrought-Iron Expanding Frame.

A first-class, strong, but light implement, adapted for a variety of purposes as a horse hoe; the front teeth stir and loosen the soil, while the back hoe cuts off all weeds and throws the soil towards the plants, much or little, according to the angle at which the operator sets it; it can also be set as a furrower, or by reversing the hoes, may be used for covering; it is supplied with three extra cultivator teeth which can replace the hoe steels when desired, thus making an unsurpassed plain cultivator.

Price, complete with wheel, as shown in the cut, $4.50.

HENDERSON'S 3=Tooth Garden Cultivator.

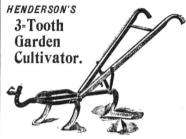

A neat, light, handy, strong, three-tooth Cultivator, low in the frame, with adjustable handles and hollow steel standard ; it is sent out with three 3-inch teeth and a 15-inch sweep. It is built specially for work in narrow rows and in the vegetable garden or where rather close cultivation is practiced and a light tool wanted.

Price, $3.00 ; or with wheel, $3.50.

THE PLANET, JR., Beet Cultivator.

Has ten teeth of an entirely new shape; they enter the earth easily, loosen the soil thoroughly, yet leave it level and without throwing dirt on the plants; the irrigating tooth follows and works like a charm; can be removed when desired. The cultivator is fitted with lever expander, lever wheel and depth regulator.

Price, $10.00.

THE PLANET, JR., 12=TOOTH

Harrow, Cultivator and Pulverizer.

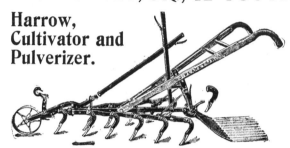

This thorough working and convenient tool has rapidly grown in favor among strawberry growers, market gardeners, truckers and farmers. This is because the twelve chisel-shape teeth do such capital work, without throwing earth on small plants, and because the tool is so convenient, durable and strong, the combination of teeth and pulverizer leaving the ground in the finest condition. The pulverizer also enables the operator to set the tool, in conjunction with the wheel, to any exact depth desired, making delicate work not only possible, but very easy. It is invaluable in narrow rows and delicate work in market gardens and close work on the farm, and is fine for preparing seed beds. The implement is fitted with lever wheel and lever expander, so the depth or width can be changed instantly while in operation; it contracts to 12 inches and expands to 32 inches. The handles can be raised or lowered to suit the driver.

Price, complete, $8.00.

The Strawberry Runner Cutter Attachment.

Is a 10-inch flat steel disc, mounted on an outrigger and attached to the 12-tooth harrow on the right side. It has a caster action, so as to follow the harrow easily, and is adjustable for depth of cut, and sidewise, and is provided with a leaf guard which is also adjustable. The guard lifts the leaves to avoid trimming them off, and thus reducing the strength of the plants.

Price, $1.50 extra.

Flat=Tooth Cultivator.

For Orchards, Orange Groves, etc.

This Cultivator is particularly intended for orange growers, orchards, cotton fields and wide shallow cultivation generally ; it can, of course, be set to run deeper when desired ; it is a first-class weed and thistle exterminator ; the frame can be expanded or contracted to suit the width of row.

Price, with wheel and three 8-inch sweeps, $4.50 ; with 10-inch sweeps, $4.75 ; with 12-inch sweeps, $5.00.

HENDERSON'S ROOT AND SUB-SOIL Cultivator.

A fine implement for deep cultivation of root crops, such as mangels, carrots, etc.

Price, $5.00.

The Planet, Jr., New 9=Tooth Cultivator and Horse Hoe.

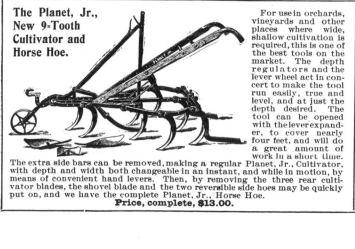

For use in orchards, vineyards and other places where wide, shallow cultivation is required, this is one of the best tools on the market. The depth regulators and the lever wheel act in concert to make the tool run easily, true and level, and at just the depth desired. The tool can be opened with the lever expander, to cover nearly four feet, and will do a great amount of work in a short time.

The extra side bars can be removed, making a regular Planet, Jr., Cultivator, with depth and width both changeable in an instant, and while in motion, by means of convenient hand levers. Then, by removing the three rear cultivator blades, the shovel blade and the two reversible side hoes may be quickly put on, and we have the complete Planet, Jr., Horse Hoe.

Price, complete, $13.00.

Grape and Berry Hoe.

CULTIVATOR ATTACHMENT.

For one horse ; adjustable to any width of row.

A wonderful labor saver in the culture of grapes and berries ; it not only does the work quickly, but often better than can be done by hand; it will take out all grass and weeds that remain under the wires and around vines and posts,and will thoroughly stir the soil close to the vine. The Hoe is guided in and out around post and vine by the Disc Caster Wheel, to which handle is attached. The horse is hitched to one side of pole, with plenty of room for plow to work under the vines or bushes, and without injury to them from horse or whiffletree. As the soil can be thrown either towards or away from the vines, the Grape Hoe can be used in grape culture when the foliage is at full growth without damage and when it is impossible to use a plow or any other tool. Constant stirring of the soil under the vine conserves moisture and improves quantity and quality of grapes.

Price, $12.00 ; or with cultivator attachment, which greatly adds to the variety and efficiency of work, $17.00.

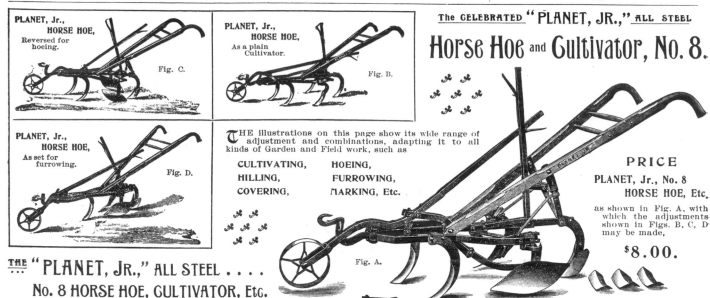

PLANET, Jr., HORSE HOE, Reversed for hoeing. *Fig. C.*

PLANET, Jr., HORSE HOE, As a plain Cultivator. *Fig. B.*

PLANET, Jr., HORSE HOE, As set for furrowing. *Fig. D.*

The CELEBRATED "PLANET, JR.," ALL STEEL

Horse Hoe and Cultivator, No. 8.

THE illustrations on this page show its wide range of adjustment and combinations, adapting it to all kinds of Garden and Field work, such as

CULTIVATING,   HOEING,
HILLING,   FURROWING,
COVERING,   MARKING, Etc.

Fig. A.

PRICE

PLANET, Jr., No. 8 HORSE HOE, Etc.

as shown in Fig. A, with which the adjustments shown in Figs. B, C, D may be made,

$8.00.

THE "PLANET, JR.," ALL STEEL . . . . No. 8 HORSE HOE, CULTIVATOR, Etc.

THIS grand implement is without a peer as a labor saver, and no progressive gardener or farmer can afford to be without it, and after once trying it would not do without it for ten times its cost. It is no ordinary, heavy, clumsy cultivator, but an implement made scientifically correct and *entirely of steel*. The patent frame is extra long and two inches higher that other cultivators, the standards being hollow, giving great strength and remarkably light weight; in consequence, it is easy for both man and horse to work. The cultivator teeth, hoes, etc., are of the finest possible quality of case-hardened, polished steel that will retain an edge; they always scour bright, and will not clog in the stickiest soil; besides they are so correctly shaped and curved that the operator can work closer up to plants without danger than with any other horse implement made, thereby entirely doing away with the usual finishing up with a hand hoe. As set in Fig. A the two forward cultivator teeth loosen the soil to any desired depth; the hoes follow, cutting off weeds below the surface and throw much or little dirt to the rows as desired, this being quickly arranged by setting the hoes at the desired angle from the plant to the centre of the row, as the hoe standards are on an adjustable swivel. Hoeing away from small plants is done by swiveling the hoe standards entirely around, as in Fig. C. **Plain cultivating** is done by replacing the three back hoes with cultivator teeth, as in Fig. B. **Furrowing** is done by arranging the tool as in Fig. D. **The Lever Expander** enables the operator instantly to open or shut the implement and lock it to suit any width of row, and can be worked while the implement is in motion, so that perfect work can be done even in irregular rows. **The Lever Wheel** enables the machine to be instantly raised or lowered and locked at any desired height while the implement is in motion, and is very convenient for changing the depth in variable soils and for raising the teeth out for turning at the end of the rows. **The Depth Regulator** is worked and set simultaneously with the wheel by the lever-bar; by this arrangement the implement runs steadily and the depth can be gauged instantly and to a nicety while the horse is in motion. **The Handles** can be quickly set to either side, and the height can be changed to suit the operator. **For Shipment** it "knocks down," reducing freight; *weight, 83 lbs.* **Price,** as shown in Fig. A, **$8.00.**

EXTRA ATTACHMENTS FOR THE Planet, Jr., No. 8 Horse Hoe, Etc. DESCRIBED ABOVE.

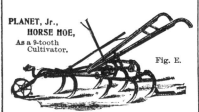

PLANET, Jr., HORSE HOE, As a 9-tooth Cultivator. *Fig. E.*

Nine-tooth Attachment. (Fig. E.) Two extra side bars with two teeth each. This makes the regular Horse Hoe into a complete Nine-tooth Machine, which leaves the ground in the finest possible condition and no furrows. Price, including depth regulators, $5.00.

PLANET, Jr., HORSE HOE, With Sweeps. *Fig. F.*

Sweeps. (Fig. F.) For flat cultivation in orange groves, cotton, peanuts, beans, garden vegetables, strawberry, quackgrass, allies, etc. Made in four sizes. Price, each, 8-inch, 30c.; 10-inch, 35c.; 12-inch, 40c.; 15-inch, 45c.

PLANET, Jr., HORSE HOE, As used for covering. *Fig. G.*

Roller. (Fig. G.) Valuable in covering corn, potatoes, etc. Price, $1.50.

Rake. A cheaper device for covering; much better in sticky land. Price, 80c.

PLANET, Jr., HORSE HOE, As a Coverer and Ridger. *Fig. H.*

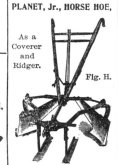

Ridging Steels. (Fig. H.) Used either for plowing away and making up beds, or covering and drawing the earth into ridges. They have adjustable wings, which can be set to make a low ridge or a high one, wide or narrow. Price, $1.10 each.

PLANET, Jr., HORSE HOE, Fitted with Vine Lifter. *Fig. J.*

Vine Lifter. (Fig. J.) Useful in sweet potatoes, tobacco, beans, peas, or any crop that needs working beneath the leaves or branches, or for crops that have fallen. Price, $1.40.

Runner Attachment. (Shown in Figs. G, H and K.) Very useful in covering and fine work, and for all work where it is desirable to straddle the row. Width can be varied from 6 to 24 inches, and the height so the tools can be worked to any depth. Price, $1.60.

Furrowing Steels. Three sizes, viz.: 10-inch, without wings, 80c.; 15-inch, without wings, $1.25; and 20-inch, with wings, which are adjustable from nearly perpendicular to nearly horizontal, or can be removed altogether (*see fig. K*), $1.75.

Marking Attachment. (Fig. K.) New pattern. Makes a wide, clean mark. Price, $2.50.

As a Furrower and Marker.

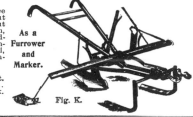

Fig. K.

The Greatest Labor Savers of the Age! ∴ ∴ A Pronounced Success!

BREED'S WEEDERS

For Horse and Hand Power.

BREED'S SULKY WEEDER.

BREED'S ONE-HORSE WALKING WEEDER.

These comparatively new implements for the farmer and gardener, though skeptically received by soil tillers when first introduced, have proved to be one of the greatest labor-saving inventions of modern times. No one who uses them intelligently would part with them for many times their cost, if they could not get another. Just think of it—if used when the weeds are small, just above the surface, one man or boy and horse can weed most thoroughly, better than by any other method, from *25 to 30 acres* a day with the No. 2, or *15 to 18 acres* a day with the No. 3, or 10 to 15 acres with the No. 4 or 5 Weeders. It seems paradoxical that an inanimate implement would know which plant to weed out and which to leave undisturbed in the soil—well, it doesn't know—but it distinguishes between them just the same. The secret is this: The seeds of crops are planted one inch or more below the surface, and their roots penetrate still deeper, while nearly all weed seeds germinate in the upper half-inch of soil, and as the supple Weeder fingers work only in the upper inch they destroy all of the weeds, if used in time, without disturbing the crop. The illustration below gives a good idea of the principle. The "Breed Weeders" are highly commended by prominent agriculturists and experiment stations, and copies of their letters or reports are printed in our special "*Breed Weeder Circular*," which will be mailed on application. This circular also gives full details as to how and when to use them.

No. 4 is our 8-ft. sectional-head Weeder with central section 2½ ft. long. This is our best all-around general-purpose Weeder, good anywhere and everywhere in all crops, even in small vegetables (except onions, unless it is in the hands of a careful man). The great advantage of this Weeder is in its having removable ends, thus permitting the use of the central section between the rows when the crops are so tall that a full-length Weeder could not be used.

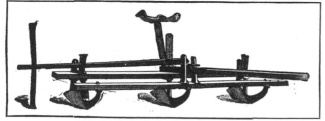

It pulls the weeds and leaves the crop.
Young weeds root only in the upper inch of soil.
Crops root below 1½ inches of the surface.

PRICES AND SIZES.

No. 2.	Two-horse Sulky Weeder, 12½-ft. head,	$37.00
No. 3.	One-horse Sulky Weeder, 8-ft. head,	30.00
No. 4.	One-horse Walking Weeder, 8-ft. head,	14.00
	The best general-purpose Weeder. See description of No. 4, above.	
No. 5.	One-horse Walking Weeder, 8-ft. head,	13.00
No. 6.	One-horse Walking Weeder, 2½ ft. head,	7.50
No. 8.	Man-power Weeder on Wheels, for onions and vegetables, head 2 ft. 9 in. long,	10.00

BREED'S MAN-POWER WEEDER.

THE PLANET, JR., DOUBLE CELERY HILLER.

Indispensable in celery growing. The leaf lifters are valuable, especially for first workings, and are adjustable in height. With them some of our best and largest growers claim to hill their celery *without any handling whatever.* It is fitted with lever wheel and lever expander. The hilling blades are 43 inches long, and are adjustable in width or height as wanted; the double machine works all rows up to four feet apart. Before hilling the soil should be thoroughly loosened with a horse hoe. **Price, $11.00.**

PLANET, JR., DOUBLE CELERY HILLER.

THE PLANET, JR., SINGLE CELERY HILLER.

The Single Hiller runs lighter than the Double, and throws rather higher; it works any width rows. The lever expander enables the operator to throw much or little earth, as desired; the height can also be regulated. Where market gardeners have their celery planted close, and first bleach every other row, and having marketed that, then wish to bleach the remaining rows, it is "just the thing."

It is also excellent help when burying the crop. **Price, $9.50.**

IMPROVED STEEL WING MARKER.

FOR POTATO AND CORN GROUND.

The long runners make a very true and even mark, which cannot be thrown out by stiff sod or stones, a difficulty with markers having short teeth. It can be adjusted to width and depth, as desired. The wearing parts of the runners are chilled iron, the wings are steel, and the frame of oak.

— PRICES. —

No. A.	For one horse—two runners, gauge and shafts (no seat)	$8.50
No. B.	For two horses—three runners, gauge, pole and seat	10.00
	Or with wheels	11.00

——COVERER AND HILLER.——

FOR POTATOES, CORN AND OTHER CROPS PLANTED IN FURROWS.

A very excellent implement, not only adapted to covering the seed, but is valuable for hilling up. The frame is adjustable to any width of row.

— PRICES. —

No. 1.	Wings of polished steel	$11.00
No. 2.	Wings of smooth, chilled iron	9.00

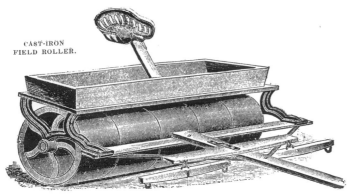

CAST-IRON
FIELD ROLLER.

CAST=IRON FIELD ROLLERS.

Furnished with either Pole or Shafts.

Sections.	Diameter.	Face.	Total Length.	Weight, About.	Manufr.'s Price.	P. H. & Co.'s Net Price.
4	20 inches.	12 inches.	4 feet.	850 lbs.	$45.00	$36.00
5	20 "	12 "	5 "	1,000 "	50.00	40.00
4	24 "	12 "	4 "	1,100 "	55.00	44.00
5	24 "	12 "	5 "	1,250 "	60.00	48.00
6	24 "	12 "	6 "	1,500 "	70.00	56.00
4	28 "	12 "	4 "	1,150 "	60.00	48.00
5	28 "	12 "	5 "	1,400 "	70.00	56.00
6	28 "	12 "	6 "	1,600 "	80.00	64.00
Whiffletrees and Neck Yoke				extra,	3.00	2.50
Scrapers				"	6.00	5.25

Cyclone Pulverizer.

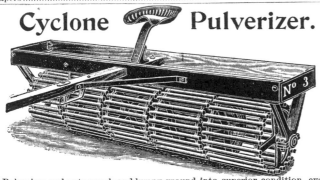

Pulverizes and puts rough and lumpy ground into superior condition, even when other implements cannot do it. It breaks the lumps while ordinary smooth surface rollers merely press them even with the surface. By its peculiar construction it leaves the surface of the ground full of indentations or depressions, which will greatly aid the absorption of water by the soil, while the old style smooth roller leaves a hard surface on top, which after a rain will bake and prevent the growth of plant. The drums and crusher bars are solid iron, which will not bend or break. Each section is 18 inches long and 18 inches in diameter. Easily taken apart for transportation and storage. The weight is about right for general work, but may be loaded to any desired weight for more severe work. The seat is on a steel spring, is comfortable, and prevents any bearing down on the horses' necks. The machine turns easily when turning a corner, when one-half the rolls turn backward, the others forward, thus allowing a corner to be turned easily and without sliding. Those using them agree that once going over with this pulverizer is better than three or four times with the harrow and smooth roller. It runs light, has no side draught and puts no weight on horses' necks. Two horses can easily operate it the entire day. One or more rolls may be removed, thus leaving the desired space in the centre, and when so arranged becomes an implement which is unequalled for rolling corn ground after plants are up. Made in four sizes, as follows:

No. 1, 4½ feet, in 3 sections, 1 horse, weight 550 lbs.........................$25.00
" 2, 6 " " 4 " 2 " " 700 " 28.00
" 3, 7½ " " 5 " 2 " " 850 " 33.00
" 4, 9 " " 6 " 2 " " 1,000 " 38.00

...OAK STAVE FIELD ROLLER.

Made of oak timber, well seasoned, 8 feet long, 28 inches in diameter. Weight, 700 lbs. Two sections, with 1¼-inch steel shafts, running through from outside brackets, on which rolls revolve independently of each other, heads tenoned into the wood and held in place by 3 long rods on each side. Chilled iron bearings; has hitch midway between tongue and ground. Price, $30.

Steel One=Horse Lawn Rollers.

Especially designed to smooth and keep in perfection

LAWNS, DRIVES, WALKS AND GROUNDS OF GOLF, TENNIS, ATHLETIC AND COUNTRY CLUBS AND GENTLEMEN'S PLACES.

The cylinders are of rolled steel, supported by hard wood felloes. The axles of machinery steel, polished work in lathe-bored journals lined with anti - friction metal, the result is a perfectly even surface and little or no noise. Each of these rollers can be weighted to three times the weights as given above, or to the practical draught capacity of a horse on sod. We also furnish an attachment for sowing seed or scattering bone dust and ground plaster. This is a very valuable adjunct, as lawns should be repaired and fertilized early in the spring, and if the seed and fertilizer are left loose on the surface, much is washed away.

Number.	Section.	Track.	Diameter.	Average Weight.	Price, Rollers Alone	Price, with Sowing Attachment.
2-L.	2	4 feet.	24 inches.	500 lbs.	$25.00	$34.00
1-L.	2	4 ft. 8 in.	30 "	600 "	40.00	49.50
0-L.	3	5 feet.	24 "	550 "	30.00	40.00
3-L.	3	6 "	24 "	600 "	35.00	45.00

SUPERIOR STEEL FIELD ROLLERS.

Furnished either with or without Seeder.

Cushioned rims, and will, therefore, stand hard usage on rough and stony land, of light draught and turns with the utmost ease. Axle of polished steel, babbitted and self-oiling bearings, drums revolving independently on the shaft when turning corners; practically indestructible; a hardened rolled steel rim close riveted at the edges. Substantial hard wood weight box; by this means the weight can be increased to the draught capacity of any ordinary team. It can be used lighter than the lightest, or heavier than the heaviest; all equally well, thus suiting time and crop, and thus properly fitting the varying requirements.

The **Seeder Attachment** sows grains, grasses and dry ground fertilizers with accuracy. Speed and outflow may be regulated. The seed, as dropped, is pressed into the soil with no opportunity for bunching or pocketing or for blowing away.

PRICES.

Number.	Sections.	Total Length.	Diameter.	Average Weight.	Price, Plain.	Price, with Seeder.
3	3	6 feet.	24 inches.	550 lbs.	$30.00	$40.00
7	3	7 "	24 "	650 "	35.00	46.00
4	3	8 "	25 "	750 "	40.00	52.00
5	3	6 "	30 "	650 "	40.00	50.00
9	3	7 "	30 "	750 "	45.00	56.00
6	3	8 "	30 "	850 "	50.00	62.00

"Straddle Crop" Steel Roller.

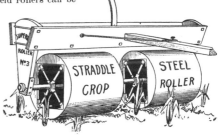

Any of our superior steel field rollers can be transformed into straddle crop rollers by removing one section; the remaining two sections can be shifted and fastened on the axle to adapt them to rows. This machine rolls both sides of a growing crop. It repairs faulty ground preparation, preserves for the plants practically all of the moistures from row to row, reduces shovel tracks, crushes clods, and gives the best results of level cultivation.

KEMP'S MANURE SPREADER.

Kemp's Manure Spreader
with Drill Attachment.

KEMP'S MANURE SPREADER **It** *spreads manure evenly. breaks it up finely. makes manure go further.*

THERE is no class or kind of farm work that involves more slave-like drudgery than the hauling and spreading of manure as it is ordinarily done. This is one of the operations on the farm that is held in awe and dread, and this is the case to such an extent that those farmers who have large quantities of manure to distribute encounter difficulty in employing labor for that purpose.

Every farmer knows that fine manure will produce better results than coarse manure; it is more valuable when fine because in that condition it approaches more nearly the solution—the actual form of plant food. In the past it has been almost impossible to reduce ordinary barnyard manure to this condition by hand labor, and when by chance, manure was reduced to a reasonably fine condition it was impossible to spread it evenly upon the ground so as to produce the best results.

With a Kemp Manure Spreader all of these difficulties are overcome. It is a complete machine, constructed upon correct mechanical and scientific principles, the sole object of which is to break up and make fine all kinds of manure and spread it evenly upon the land in any desired quantity per acre; it will spread very coarse manure, corn-stalks, clover haulm, chaff, sawdust, or such other substances as are intended for mulching; it will also spread lime, wood ashes, coal ashes, marl, plaster, salt or any substance intended for use as manure.

It will also distribute manure, lime, ashes, cotton seed, cotton seed hulls and commercial fertilizers in trenches or drills, better, quicker and more economical than can possibly be done by hand.

The illustration above gives a very correct idea of the spreader as it appears at work in the field. The apron, which constitutes the bottom of the machine, moves slowly backward, propelled by the gear wheels, thus the manure is brought into contact with the rapidly revolving cylinder, which is studded with 115 steel teeth which, aided by the spring rake above the cylinder, completely fines, pulverizes and disintegrates the manure, which falls upon the ground in an even, steady stream, covering every particle of the surface. This even distribution and pulverizing of the manure, according to the estimation of those who are most competent to judge—men who have used the Spreader for ten years or more—increases its value from 50 per cent. upwards. With the horses at a smart walk, a load of manure may be distributed with this machine in from three to five minutes, depending upon the amount the machine is set to apply to the acre.

The illustration shown on the right represents the Spreader distributing composted manure direct to the drill, before the seed is planted. Market gardeners, potato growers, farmers, small fruit growers, tobacco and cotton planters, find this machine, with the drill attachment, the one thing needful to enable them to double the value of the product of their acres. The drill attachment will distribute manure in a single drill or in two drills ranging in width from 1 foot to 5½ feet apart; it will cover ten times as much ground with a given amount of manure and produce equally as good or better results.

PRICES:

Size No. 1, cap'ty 30 bush. (suitable for small teams or hilly lands), **$100.00**	Hood (for spreading lime, ashes, and for use in windy weather), extra **$5.00**
Size No. 2, cap'ty 40 bush. (suitable for good team and rolling lands), 105.00	Brake for hilly sections, extra 5.00
Size No. 3, cap'ty 50 bush. (suitable for good team and level lands), 110.00	Three-horse evener, extra 5.00
Drill attachment for leaving manure in rows, extra 8.00	Slow feed for lime, ashes, etc., extra 1.50

Whiffletrees, neck yoke and two-horse evener furnished free with machines.

THE STEVENS FERTILIZER SOWER.

THE LATEST IMPROVED AND BEST MACHINE FOR SOWING ALL KINDS OF ARTIFICIAL FERTILIZER,

such as bone, phosphates, cotton seed meal, ashes, lime, pomace, etc., either damp or dry, in drills or broadcast.

Unequalled for top-dressing grass lands or lawns.

THE
STEVENS
FERTILIZER
SOWER.

It distributes any quantity desired from 200 lbs. to 4,000 lbs. per acre, and five times as fast as can be done by hand; it is as great an improvement over hand spreading as the mowing machine over the scythe; it sows very evenly, no clogging, with strings, sticks, lumps or stones, and there are no wheels to become gummed up. This machine is set in motion by a lever operated with the foot, so simple that a boy can run it; it is light of draft, even in soft land, as the wheels have tires four inches wide, also a ratchet hub which allows the turning of corners or completely around, while in motion or to back, without throwing out of gear. It is low down and easy to fill. All parts are thoroughly constructed and this Sower will last a lifetime. Adjustment is perfect, simple and strong, a great improvement over the complicated adjustment used in other sowers. It broadcasts 5 feet 10 inches wide. Combined pole and shafts with neck yoke and whiffletrees, complete for hitching on either two horses or one, will be sent with all machines, unless ordered with shafts only. In making this combination, bent or crooked shafts are used, overcoming the great objection made to straight shaft combination. **Price, $48.00.**

We guarantee perfect satisfaction on every machine.

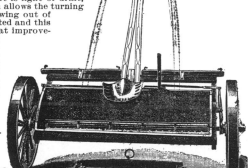

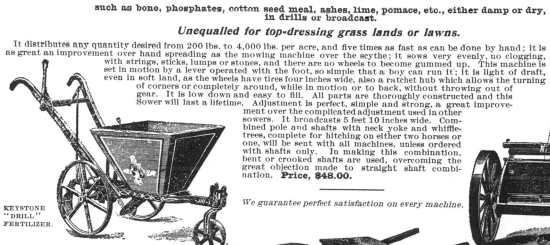

KEYSTONE
"DRILL"
FERTILIZER.

Keystone "Drill" Fertilizer One Horse Power.

For truckers, farmers, cotton and tobacco planters. Superior to any machine of its class on the market. Positively force feed will sow 200 to 600 lbs. to the acre with perfect regularity. Will sow all kinds of fertilizer. Box holds a 200-lb. bag of fertilizer; improved swivel clevis, all steel frame and cold rolled steel shafts. The machine is thrown out of gear and the shovel raised by one operation. Strong and handsomely finished. **Price, $12.00.**

ADVANCE MAN-POWER FERTILIZER DRILL.

"Advance" Man-Power Fertilizer Drill.

The best low-priced distributer on the market, and would call attention to its distributing disc, of galvanized iron, which can neither break nor rust; also to the wrought-iron wheel, light and strong. A shut-off, to prevent the escape of fertilizer when wheeling around end of row, is a late improvement. A first-class tool for the drilling of peas and corn. **Price, $7.00.**

ECLIPSE One-Row CORN PLANTER

AND

Fertilizer Distributer.

Plants field or ensilage corn, beans, peas or beet seeds in hills, drills or checks; it will drop 6, 12, 24 or 36 inches apart. By scattering pumpkin seeds in with the fertilizer in filling, hopper will place seeds in hills when planting corn. It accurately distributes from 50 to 450 lbs., as desired, of commercial fertilizer, wet or dry, in the rows or hills, first drawing moist earth on the seed before the fertilizer is dropped each side of the seed, and all is covered to a uniform depth. The positive check row attachment will check in seed and fertilizer any desired distance, regardless of irregularity; it is so constructed that the dropper cannot deposit the seed for a hill, in a tube, until the one already in has been let out; the seed dropper and valves, operated by the planter, are controlled by a touch of the finger; check row attachment, as shown in the cut, should only be attached when planting in checks. **Price complete, $25.00.**

"KING OF THE CORNFIELD" Two-Row

Corn, Bean and Pea Planter and Fertilizer Distributer.

Does perfect work, not only in planting corn in hills or drills but does equally as good work planting peas, beans and all similar seeds, putting in the fertilizer as well in any desired quantity, or the fertilizer may be stopped off altogether, if desired. Another great feature in this machine: corn and beans may be planted at the same time in the same row, the beans half way between the corn or 4½ or 9 inches from corn, as desired. The fertilizer may be put with the corn and left out of the beans, if desired to do so. Pumpkin can also be planted in connection with corn and beans at the same time; plants rows 30 and 36 inches apart. **Sizes :** 24-inch wheel, plants 4½, 9, 18, 36 inches, **$50.00** ; 20-inch wheel, plants 3¾, 7½, 15, 30 inches, **$49.00.**

MILLER'S BEAN PLANTER.

It plants in hills 14 inches on the row; the distance can be changed to suit the operator. Distance between rows is 28 inches, but can be changed when desired. The droppers are adjustable to plant all sizes and varieties of beans and do not crush or crack the seed, and will plant either corn or beans. Its marking and covering device is an improvement over all other planters in use, as it clears away all stone and clay or dry lumps which are a hindrance to the crop, and allows nothing but fine and moist dirt to fall back upon the seed, thus making a march of the crop, coming up evenly. A phosphate attachment is furnished with the planter when ordered. The wheels perform the marking or guiding by returning upon the wheel mark. **Price,** 2-horse planter, **$35.00;** or with Fertilizer Attachment, **$45.00.**

THE "HENDERSON"

Corn Planter and Fertilizer Distributer.

The best Corn Planter and Fertilizer Distributer in the world.

It cracks no grains and will plant from ten to twelve acres of corn per day, dropping kernels in drills or in hills at any desired distance apart and sowing at the same time, if needed, any kind of pulverized fertilizer; it is simple enough for the most ordinary laborer to use without getting it out of order. Each machine is furnished with four dropping rings and pinions to regulate the number of kernels and distance apart of planting. In addition to these four rings, we furnish rings at 25 cents a piece extra to plant peas, beans and other seeds, which, by the arrangement of the holes in the rings and the different pinions we furnish, can be distributed in any way that a farmer could wish. Some use it entirely for planting ensilage, others for planting peas, beans, etc. The phosphate attachment is perfect. The phosphate is accurately distributed and with absolute safety from injury to the seed; at the same time it may be placed near enough so that the young plants may at once receive the intended nourishment to give them an early start.

Steel frame..$14.00
"   "   with phosphate attachment.................... 18.00

MILLER'S BEAN HARVESTER.

This Harvester surpasses any other machine of its kind. By means of the rods on the flexible, rolling dividers, the vines are gathered and brought together into a windrow at the rear of the machine, free from roots and dirt. The adjustment of the machine is simple, and may be operated, as to depth, easily from the seat by means of the levers. The drive-wheels are provided with ribs in the centre of the rim, which prevents the machine from slipping sideways, either on level ground or on hillsides. The guards in advance of the drive-wheels remove all loose stones from their paths, which would otherwise raise the machine and be a hindrance in the performance of its work. The pole has a tilt-lever that will give different pitch to the machine. It also is provided with a sway-lever that shifts the draught and the pole from right to left, which makes it the best machine for side-hills and difficult places. **Price, $30.00.**

THE "BEST" CORN HARVESTER.

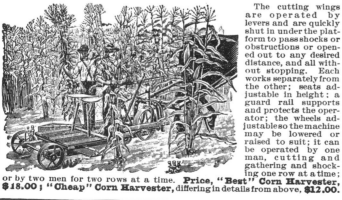

The cutting wings are operated by levers and are quickly shut in under the platform to pass shocks or obstructions or opened out to any desired distance, and all without stopping. Each works separately from the other; seats adjustable in height; a guard rail supports and protects the operator; the wheels adjustable so the machine may be lowered or raised to suit; it can be operated by one man, cutting and gathering and shocking one row at a time, or by two men for two rows at a time. **Price, "Best" Corn Harvester, $18.00; "Cheap" Corn Harvester,** differing in details from above, **$12.00.**

CORN AND COTTON STALK CUTTER.

It is balanced so there is no pressure at any time nor jerking motion on the horses' necks. The hooks for raking stalks in line are automatically arranged so as to raise with the cylinder. The springs on sides that give pressure to cylinders can be given more or less tension as desired. The knives are set at that peculiar angle which insures thoroughly cutting the stalks. With the lever you can raise the knives off the ground and can also regulate, to a certain extent, downward pressure, insuring cutting of stalks in unfavorable weather. The wheels and seat are high, so the driver is well out of the dust.

Price, Single-Row Stalk Cutter, for two horses, $35.00
"   Double   "   "   "   three "   54.00

Aspinwall Riding Corn and Bean Planter.

Plants Beans, Corn, Peas and Distributes Fertilizer.

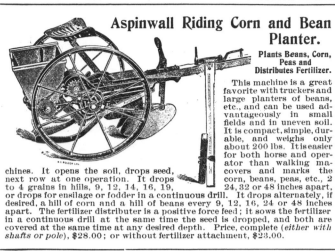

This machine is a great favorite with truckers and large planters of beans, etc., and can be used advantageously in small fields and in uneven soil. It is compact, simple, durable, and weighs only about 200 lbs. It is easier for both horse and operator than walking machines. It opens the soil, drops seed, covers and marks the corn, beans, peas, etc., 2 next row at one operation. It drops 24, 32 or 48 inches apart, to 4 grains in hills, 9, 12, 14, 16, 19, or drops for ensilage or fodder in a continuous drill. It drops alternately, if desired, a hill of corn and a hill of beans every 9, 12, 16, 24 or 48 inches apart. The fertilizer distributer is a positive force feed; it sows the fertilizer in a continuous drill at the same time the seed is dropped, and both are covered at the same time at any desired depth. Price, complete (*either with shafts or pole*), $28.00; or without fertilizer attachment, $23.00.

Buckeye One=Horse Grain Drill.

FOR DRILLING AMONG STANDING CORN.

A first-class machine; is mounted on three wheels, the back one being a large caster wheel, allowing short, quick turns at ends of rows. The dragbars work independently of each other, allowing any one of the five hoes to pass an obstruction without raising the others out of the ground. There is a device for spreading the hoes to suit the space between the rows. The grain feed is a positive force feed, and will sow all kinds of grain evenly, and may be set to sow any quantity desired. The machine is neat, light, strong and works admirably. Price, $25.00 each; or combined with fertilizer sowing attachment, $30.00.

Cahoon's Improved Broadcast Seed Sower

Will sow all kinds of grass and grain seeds from 4 to 8 acres per hour at a common walking gait. Heavy seeds, such as wheat, it will throw 40 feet (20 feet each side of the operator); lighter seeds, of course, will not be thrown so far. Price, $3.75.

CAHOON'S BROADCAST SEEDER.

Low Down Force Feed PENNSYLVANIA Grain Drill.

Furnished either with or without Phosphate Attachment, and with either Pin or Spring Hoes.

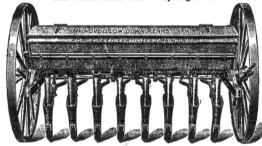

This drill is unquestionably the simplest, most accurate and lightest-running drill made. The bottom of the hopper is only about 3 feet from the ground, making it easy to fill, balancing the machine perfectly and reducing draft. The grain feed is a positive force feed, yet does not crack or injure the seed in the slightest The quantity sown is quickly regulated by a lever. The land measure or clock is correct, and quickly adjusted before commencing the day's work. The hoes can be instantly changed by a lever, even while the machine is in motion, to run either straight or zigzag. The grass seeder is in the rear and distributes the seed behind the hoes, though it can be furnished to sow in front of hoes if so ordered. The phosphate attachment has a positive force feed, not only in name but in fact, and distributes fertilizer to perfection; driving wheels large, wide face, and everything about the machine of the highest class. Price, 8-tube, 8-inch, complete with grain, grass seed and fertilizer attachments, $75.00.

THE HAND SEED SOWER.

FOR SOWING VEGETABLE SEEDS IN GARDENS AND HOT BEDS.

It will open the drill, sow and cover beet, cabbage, carrot, celery, lettuce, radish, turnip and all such seeds with perfect regularity. It sows much more evenly and ten times as rapidly as by hand. The quantity to be sown can quickly be regulated, and also the depth. It is the only drill made for sowing in hot beds. Will sow a small packet of seeds as well as larger bulks. Simple, easily understood, and cannot get out of order. Price, $1.50.

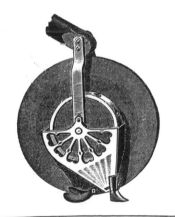

The Strowbridge Broadcast Seeder.

Attaches to any ordinary farm wagon, and sows to perfection grass and grain seeds of all kinds and in any desired quantity per acre 50 to 100 acres a day; it also distributes all dry commercial fertilizers, also ashes, land plaster, lime, etc. It has a positive force feed and an adjustment opening the holes from ¼ to 1½ inches, which allows the sowing of seeds of any size. The flow can be shut off anywhere, either on one side or the other or altogether, as required; it is instantly thrown in or out of gear.

Price, complete to attach to wagon, $12.00.

Thompson's Original Wheelbarrow Grass Seeder.

It cannot clog. The wheel is large, 33 inches, to give the greatest force with the least exertion. It balances perfectly. It sows evenly, the whole length of the hopper. The quantity of seed is easily regulated.

No. 1. 14 ft. single hopper, for clover, timothy, Hungarian, millet, alfalfa, flax and rape, from 2 to 12 qts. per acre. Price, $10.00.

No. 2. 14 ft. double hopper, sows the same as No. 1, from one side of the hopper; from the other side it sows orchard, red top and similar grass seeds in any quantity desired per acre. It can be used for either purpose, by reversing upon the wheelbarrow shafts. Price, $13.00.

No. 5. 14 ft. double hopper, sows red top, orchard and similar grass seeds in any quantity per acre, and from other side of hopper all seeds like clover, timothy, Hungarian, alfalfa, etc., from 6 to 40 qts. per acre, and millet 3 to 24 qts. per acre. Price, $12.00.

No. 4 is a grain seeder with a 10 ft. hopper for broadcasting wheat, rye, oats, barley and buckwheat. Price, $11.00.

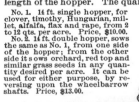

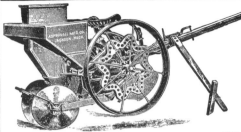

THE ASPINWALL POTATO PLANTER, WITH FERTILIZER
ATTACHMENT AND DISC COVERERS.

The fertilizing attachment is accurate, reliable and never becomes clogged; it can be attached or removed at pleasure. The quantity per acre can be gauged to suit anywhere from 100 to 1,500 lbs.

It can be thrown out of gear instantly and it distributes the fertilizer just above the seed, after sufficient earth has fallen upon the potato to prevent injury.

The New Disc Coverers in some soils are preferred by some planters, but it is not expected that they will be suitable for all localities, for it is understood that their use is limited.

The Aspinwall Potato Planter.

The Aspinwall Potato Planter has for 13 consecutive years done the most successful and satisfactory work under all the many varying conditions of planting. It is the leading machine of the world. Its splendid record, founded upon its dozen years of triumphs, places it far and away ahead of any competitor. It is in no sense an experimental machine but a demonstrated and practical success. No other machine can begin to compare to it in points of simplicity, durability, effectiveness of working qualities and economy. It performs the work of 10 men and does it more satisfactorily. It is the only real automatic planter made. If you grow potatoes in considerable quantities, you can no more afford to do without this machine than a farmer can do without his mower.

One man operates the Aspinwall Potato Planter; no other help of any kind is required. It marks the rows, opens the furrows, drops and covers the seed in one operation. It plants whole or cut seed 10, 13, 15, 17, 21 and 26 inches apart, and plants the rows the same distance apart. It plants 5 to 8 acres per day, and deposits moist or under-earth upon the seed. It plants the seed in a perfect line from 3 to 9 inches deep and covers uniformly. It plants wherever it is possible for the plow to work,

PRICES: Aspinwall Potato Planter, with either regular or disc coverers, as preferred...$65.00
Or with Fertilizer Attachment.. 75.00

Triumph Hand Corn Planter.

The only Planter containing an adjustable dropping disc.

The disc is quickly adjusted for a light or heavy seeding, and rotates similar to the disc in a Horse Planter.

The Planter is easily operated by a boy or girl.

Works in all kinds and conditions of soil, and is the only one which insures an accurate and reliable seeding.

PRICE, $2.00.

TRIUMPH CORN
PLANTER.

The Hand .. Potato .. Planter.

Walk erect, drop seed in the tube, plunge planter deep in the soil, press forward as you withdraw it and the deed is done, and the seed is left in moist soil. The planter cannot clog —the bottom of the tube is larger than the top.

PRICE, $2.50.

HAND POTATO PLANTER.

The Bemis Plant Setting Machine.

Transplants all kinds of plants at any required distance, and at the same time water is deposited at the roots and covered by dry, fine earth which cannot bake and which retains moisture underneath. It does the work far better than when done by hand. The machine is drawn by two horses and the work is done by a driver and two boys, who do the operating. It plants one row at a time and can transplant from 3 to 6 acres a day, according to the distance between plants and the skill of the droppers. The rows may be 30 inches apart or as much wider as desired. One foot apart in the rows is about as close as ordinary operators can transplant, but when expert, they can operate fast enough for celery.

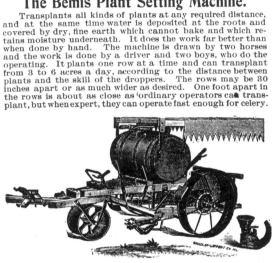

For tomato, cabbage, cauliflower, sweet potatoes, strawberries, tobacco and small nursery plants, etc., or any plants that do not require to be planted closer than 1 foot, the machine works to perfection, and for extensive truckers, etc., it will pay for itself several times in one season. Plants may be set deep or shallow, and the quantity of water to each regulated from nothing to 6 barrels an acre. Roots are not doubled up as in hand planting; plants start to grow quicker, mature more evenly, and the grower is independent of labor and has no lame back—no delay in planting on account of dry weather. Plants are set straight, allowing closer and quicker cultivation. It is easy of draft and a thoroughly good machine in every respect, and will last a lifetime.

PRICE, $70.00, or with Phosphate attachment, $80.00.

Extra.	Extra.
Potato planter.................$5.00	Check Rower, with 80 rods
Tree-planting attachment . 5.00	wire...........................$ 15.00

Aspinwall Potato Cutter.

With unerring precision it divides potatoes into quarters, halves or any number of parts. Separates the eyes and removes the seed ends. Any boy can operate it and do the work of 10 men by hand. It greatly facilitates the work of preparing the potatoes for planting.

Bed of Knives, showing manner in which potato is cut.

A series of knives in a frame-work, or table, comprises the machine. The potato is placed between a pair of jaws above the knives which, by a single stroke of the plunger, sever or dissect the potato, dividing it as has been stated.

PRICE, $10.00.

The Hoover Potato Digger.

The very best digger made. It separates the potatoes and delivers them behind the machine in a narrow, even row on cleaned ground, while the vines and weeds are carried to the side. The depth and position of rack are regulated by levers. The shovel is guaranteed against stumps and stones. Two horses operate the machine under ordinary conditions, but the digger is guaranteed against four horses if extraordinary conditions require them.

PRICE, f. o. b. factory. $110.00.

The Aspinwall Potato Sorter

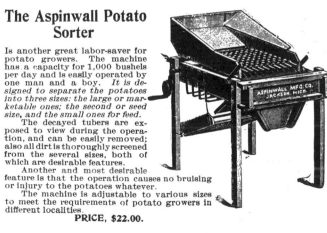

Is another great labor-saver for potato growers. The machine has a capacity for 1,000 bushels per day and is easily operated by one man and a boy. *It is designed to separate the potatoes into three sizes: the large or marketable ones; the second or seed size, and the small ones for feed.*

The decayed tubers are exposed to view during the operation, and can be easily removed; also all dirt is thoroughly screened from the several sizes, both of which are desirable features.

Another and most desirable feature is that the operation causes no bruising or injury to the potatoes whatever.

The machine is adjustable to various sizes to meet the requirements of potato growers in different localities.

PRICE, $22.00.

THE "CASE=KEELER" SEEDER.

For Hand Power.

The King of Seeding Devices.

FIG. B.

The Case-Keeler as a plain seeder.

FIG. C.

The Case-Keeler Seeder adjusts for going up or down hill, sowing equally well as on level ground.

FIG. D.

Case-Keeler, with infant plows, cultivating away from plants.

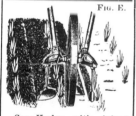

FIG. E.

Case-Keeler, with infant plows, cultivating towards the plants.

FIG. F.

Case-Keeler, with infant plows and cultivator teeth in use.

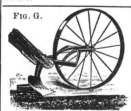

FIG. G.

Case-Keeler, with plow attached, for hilling, furrowing, covering, etc.

An entirely new machine, built on new principles, and guaranteed to do all claimed for it.

It sows in a continuous row or plants in hills, 6, 12, 18 or 36 inches apart, as desired, and as fast as a man can walk.

It sows or plants, with absolute regularity, all kinds of seeds, fine or coarse, without a break, scrape, split or bruise. It successfully handles beans, beet, cabbage, carrot, celery, corn, lettuce, onions, parsnip, peas, salsify, spinach, turnip, etc., etc. It will plant cucumber and melon, scattering the seeds in each hill six inches, if desired.

It sows or plants any desired depth, opens the furrow, drops the seed, covers it, rolls it and marks out the next row, all in one operation.

It is very durable and will last for years; the working parts, being of brass and cast zinc, will not rust.

It is very simple—cannot get out of order. The changes can be made in half a minute and will be understood by a child.

The Seeder is entirely removed while the other tools are being used, thus doing away with the great disadvantages of all other combined machines. **Price of Plain Seeder,** with handles and frame (as shown in FIG. B), **$9.00.**

Price of combined machine,

as shown in **FIG. A,** which includes two flat hoes, three cultivator teeth, two rakes and one plow,

$12.00.

(Weight, crated for shipment, 65 lbs.)

FIG. A.

FIG. H.

Case-Keeler, with rakes attached, for 1st shallow cultivation.

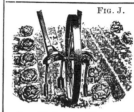

FIG. J.

Case-Keeler, with cultivator teeth attached, for 2d or deep cultivation.

FIG. K.

Case-Keeler, with flat hoes, working from the plants.

FIG. L.

Case-Keeler, with flat hoes, working towards the plants.

FIG. M.

Case-Keeler, with leaf arm, protecting plants.

The "Case-Keeler" Wheel Hoe, Cultivator and Plow,
FOR HAND POWER.

This Machine, for hoeing, cultivating and hilling, is not surpassed by any implement made. The **two flat hoes** (see FIGS. K, L and M) work close to the plants and cut off all weeds. The **three cultivator teeth** (FIG. J) follow and loosen the soil two or three inches deep and pulverize it, or the cultivator teeth and hoes can be used separately, if desired. The machine is also supplied with **two rakes** (FIG. H) for fine surface cultivating and leveling. The **two infant plows** (FIGS. D, E and F) are for throwing earth either to or from the plants, as desired, and **one large single plow** (FIG. G), which will work from three to six inches deep, and is invaluable for hilling up, furrowing, etc. These attachments can be adjusted to different widths of rows and can be operated between two rows or be adjusted to straddle and work both sides of one row.

The Leaf-arm (FIG. M) pushes aside overhanging leaves, allowing the Cultivator to "work close," even when plants are large.

The machine can be pushed at a moderate or fast walk, doing the work of six men with the customary tools, doing it better and with far greater ease. The 20-inch drive-wheel makes the machine easy to push.

Price of "Case-Keeler" Wheel Hoe, Cultivator and Plow, $8.00.

Note.—*The wheel and frame are exactly the same for both Seeder and cultivating and hoeing tools. If you purchase the Seeder alone you can afterwards add any or all of the cultivating tools, or if you purchase the Wheel Hoe, etc., you can afterwards add the Seeder, at the under-mentioned prices, viz.:*

Wheel, frame and handles	$4.00	Double infant plows	$1.00
"Case-Keeler" Seeder attachment	5.50	Cultivator teeth, set of three	.75
Plow	1.00	Onion set sower	.50
Double rakes	.75	" " " covering steels	.50
" hoes	1.00	Leaf-arm	.50

FIG. N.

Case-Keeler onion set sowing attachment makes wide drill and scatters seed. Price, without covering steels, 50c.; with covering steels, complete, $1.00.

PLANET, JR., FERTILIZER AND PEA DRILL

FOR HAND POWER.

Holds one-half bushel. The fertilizer is fed in a wide, even stream through a rear discharge mouth, which is regulated in size by a feed rod and an index at the top of the handle.

The tool is instantly thrown out of gear by the feed rod.

It will sow fertilizers evenly from 100 lbs. to 1,000 lbs. to the acre ; with material in any reasonable condition, it will not clog ; is galvanized, has no cogs, gears or stirring devices, and will give perfect satisfaction.

It sows peas in any quantity and with the greatest regularity.

PRICE, COMPLETE, $15.00.

PLANET, JR., No. 5

LARGE CAPACITY

SEEDER FOR HILLS AND ROWS.

ESPECIALLY ADAPTED FOR MARKET GARDENERS AND LARGE PRIVATE GARDENS.

This is our largest Hand Garden Seed Sower and is adapted for market gardeners' use and large hotels or private gardens. It drops all kinds of garden seeds from the finest to peas and beans at 5, 6, 8, 12, 16, 24 or 48 inches apart or in a continuous row, as desired. It opens the furrow deep or shallow, drops the seed, covers and rolls it and marks the next row at one passage. It is the latest improved and one of the best Seeders in' our list.

No cultivating or hoeing tools can be attached to this machine.

PRICE, $12.00.

THE PLANET, JR.,

COMBINED HILL DROPPING SEED AND FERTILIZER DRILL.

To meet the pressing demand for a drill which will sow seed and fertilizers, both at the same time, this combined machine is made. In addition to planting all garden seeds in rows and in hills, 4, 6, 8, 12 or 24 inches apart, it also sows the fertilizer at the same operation and with the same regularity and economy. The seed hopper holds 2 quarts and the fertilizer hopper 4 quarts.

PRICE, $16.00.

PLANET, JR., No. 4

FAMILY GARDEN

HILL DROPPING AND

CONTINUOUS ROW

SEEDER, ETC.

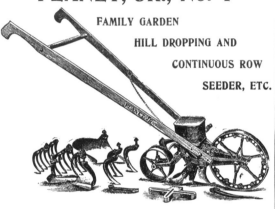

This is an unequalled machine for the family garden. It sows all garden seeds from the smallest up to peas and beans in a continuous drill, or will drop in hills 4½, 6, 9, 12, 18 or 36 inches apart. It sows with the utmost accuracy as to quantity, distance and depth, opening the furrow, dropping the seed, covering and rolling it and marking for the next row in one operation. It is instantly thrown in or out of gear—for turning, etc. The hopper holds 2 quarts of seed, but it will sow a small quantity just as well.

In addition to the Seeder, the complete machine is furnished with cultivating and hoeing tools, as shown in the cut above ; to use the latter the Sower is removed.

PRICE, COMPLETE, - $10.00.
" AS A DRILL ONLY, 7.00.

PLANET, JR., No. 4

AS A GARDEN HOE AND CULTIVATOR.

THE PLANET No. 4 (*described in column to left*) makes one of the best hand power Garden Cultivators on the market. The Seeder is entirely removed, leaving the machine light and handy for hoeing, cultivating, hilling, etc.

PLANET, JR., No. 1

COMBINED DRILL, CULTIVATOR, RAKE AND PLOW.

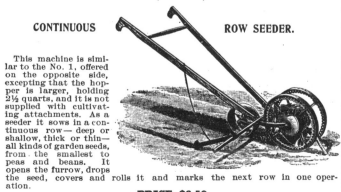

The old reliable continuous Row Seeder sows all kinds of garden seeds with great regularity — either shallow or deep, thick or thin, as desired. It opens the furrow, sows the seed, covers it, rolls it and marks the next row in one operation ; besides this it is supplied with cultivating and hoeing tools and a plow for hilling, furrowing, etc.

PRICE, COMPLETE, $9.00.

PLANET, JR., No. 2

CONTINUOUS   ROW SEEDER.

This machine is similar to the No. 1, offered on the opposite side, excepting that the hopper is larger, holding 2½ quarts, and it is not supplied with cultivating attachments. As a seeder it sows in a continuous row — deep or shallow, thick or thin— all kinds of garden seeds, from the smallest to peas and beans. It opens the furrow, drops the seed, covers and rolls it and marks the next row in one operation.

PRICE, $6.50.

PLANET, JR., No. 15 SINGLE WHEEL HOE, CULTIVATOR, RAKE AND PLOW.

Price complete as shown in the cut, $6.00.

This is the very latest and best of the Planet single wheel cultivating implements. The wheel is larger and can be raised or lowered, and can be set on one side for hoeing two sides of a row at once; the handles are changeable in height. The outfit consists of 1 land side plow for shallow plowing, furrowing, covering, hilling, etc.; 3 rakes for shallow cultivation, fining, leveling and pulverizing soil; 3 cultivator teeth for deep stirring of soil; 4 flat hoes of different widths for loosening crust and cutting off weeds, and a leaf guard—a most valuable machine.

This old favorite has always given satisfaction. It is considerably lighter than the double wheel Hoe and does grand work between rows, but does not straddle them. It is supplied with two flat hoes for cutting weeds, loosening the surface, etc.; 2 rakes for light cultivating, leveling and fining the soil; 3 cultivator teeth for stirring the soil deeply, and 1 land side plow for furrowing, hilling up and covering; also 1 leaf guard. The wheel has a broad face and can be quickly changed for depth.

PLANET, JR., No. 16 SINGLE WHEEL HOE, CULTIVATOR, RAKE AND.... PLOW.....

Price complete, as shown in the cut, $5.00.

1st Hoeing with Planet, Jr., Double Wheel Hoe.

2d Hoeing with Planet, Jr., Double Wheel Hoe.

Earthing-up with Planet, Jr., Double Wheel Hoe.

PLANET, JR., No. 11 DOUBLE WHEEL HOE, CULTIVATOR, RAKE AND PLOW.

THE LATEST AND BEST DOUBLE WHEEL CULTIVATOR, Etc.

Price complete, as shown in the cut, $8.00.

The latest improved double wheel machine, combining all of the merits of the older styles with several desirable new features. The wheels, 11 inches in diameter, can be set 9 or 11½ inches apart for narrow rows, and 4 inches apart when used as a single wheel Hoe; the frame is malleable, with a quick change device so the position of the tools may be changed without removing the nuts; the arch is 20 inches high, allowing it to straddle crops when well grown. The outfit of tools comprises the greatest variety and all of the latest designs, covering every possible need. They are as follows: 1 pair each of 4 and 6 inch hoes, 1 pair each of 3-tooth and 5-tooth rakes, 1 pair of plows, 1 pair each of narrow and wide cultivating teeth, and 1 pair of leaf guards. It is an unequalled garden implement and, although costing you more than some others, you will have the best tool of its kind manufactured. Price complete, as shown in the cut, $8.00.

Cultivating with Planet, Jr., Double Wheel Hoe.

Shallow Cultivation with Planet, Jr., Double Wheel Hoe.

Onion Harvester

ONION.... HARVESTER.

Fits any of the Planet, Jr., Double Wheel Hoes; loosens onions or sets without scattering them; fine for cutting Spinach. Price, $1.00.

PLANET, JR., $6.00 DOUBLE WHEEL HOE (No. 12) CULTIVATOR, RAKE AND PLOW.

Price complete, as shown in the cut, $6.00.

This old reliable will straddle plants 18 inches high and finish up rows from 6 to 18 inches apart at one passage. It is invaluable in all garden work, and one operator will do as much with it as several men with ordinary hoes. It is supplied with 1 pair of flat hoes for cutting weeds and loosening the crust, 1 pair of plows for hilling, furrowing, etc., 2 pairs of cultivating teeth for stirring the soil deeply, 1 pair of detachable leaf guards to lift plants and foliage, preventing injury and covering them with soil. Price, (No. 12), complete, as shown in the cut, $6.00; or with only 1 pair 6 inch flat hoes, (No. 13), Price, $4.00.

THE "HENDERSON" HAND GARDEN PLOW, FURROWER, HILLER AND SCUFFLE HOE....

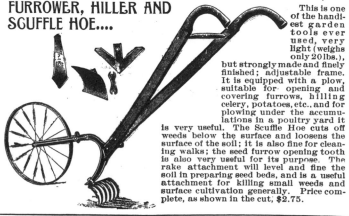

This is one of the handiest garden tools ever used, very light (weighs only 20 lbs.), but strongly made and finely finished; adjustable frame. It is equipped with a plow, suitable for opening and covering furrows, hilling celery, potatoes, etc., and for plowing under the accumulations in a poultry yard it is very useful. The Scuffle Hoe cuts off weeds below the surface and loosens the surface of the soil; it is also fine for cleaning walks; the seed furrow opening tooth is also very useful for its purpose. The rake attachment will level and fine the soil in preparing seed beds, and is a useful attachment for killing small weeds and surface cultivation generally. Price complete, as shown in the cut; $2.75.

"IRON AGE" SINGLE WHEEL HOE,

WITH

CULTIVATOR TEETH, PLOW and RAKES.

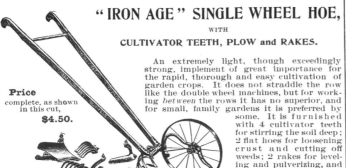

Price complete, as shown in this cut, **$4.50.**

An extremely light, though exceedingly strong, implement of great importance for the rapid, thorough and easy cultivation of garden crops. It does not straddle the row like the double wheel machines, but for working *between* the rows it has no superior, and for small, family gardens it is preferred by some. It is furnished with 4 cultivator teeth for stirring the soil deep; 2 flat hoes for loosening crust and cutting off weeds; 2 rakes for leveling and pulverizing, and one land-side plow for hilling up, furrowing, covering furrows, etc.; also, one leaf-guard, to lift leaves and vines aside, preventing injury and covering them with soil. **Price, $4.50.**

"IRON AGE" DOUBLE WHEEL HOE,

WITH

CULTIVATOR TEETH, PLOWS and RAKES.

An indispensable implement for the cultivation of garden crops. With it one person does the work of several men, and with great ease. It has four cultivator teeth for stirring the soil deeply; 1 pair of plows for plowing away from or for hilling, opening furrows and covering them after the seed has been planted; a pair of rakes for fining and leveling seed beds, weeding and shallow

Price complete, as shown in this cut, **$6.00.**

cultivating; a pair of vine lifters (detachable), valuable for lifting leaves and vines to prevent covering them with soil; a pair of flat hoes for cutting off weeds just below the surface, and loosening the crust. The machine is very light and strong, high bicycle wheels, highly arched frame of tubing, enabling it to work astride of plants twenty inches high. **Price, $6.00.**

WEEDER ATTACHMENT

FOR THE "IRON AGE"

SINGLE WHEEL HOE, Etc.

Of great utility in destroying young weeds, pulverizing and leveling the soil. etc. **Price, 75c.**

THE CUT BELOW

Shows the "Iron Age" *Double* Wheel Hoe and Combined Drill converted into a single wheel machine. We furnish the extra axle free with both of these machines, enabling any one to quickly make the change, when desired, for working between rows.

"IRON AGE" COMBINED SEED DRILL,

WITH

WHEEL HOE, CULTIVATOR, RAKE and PLOW.

Price complete, as shown in this cut, **$9.00.**

The Seed Drill is complete in itself, merely requiring to be bolted with two bolts to the wheel frame. In using the machine as a cultivator, hoe or plow, the drill attachment is taken off and put aside, so there is no wear on it nor danger of injury while being used as a cultivator. As a drill it opens the furrow, drops the seed, covers and rolls it and marks the next row in one passage. It can be set to sow shallow or deep, thick or thin, as desired, and handles all garden seeds. The discharge index is plain, convenient and reliable, and can be quickly regulated to a hair, and the flow of seed can be instantly shut off to turn rows. The agitator is a revolving brush, which will not injure seed in the slightest; will separate seeds that stick together, and makes a positive force feed. This brush is made of best bristles, and will last for a long time, and can be cheaply replaced when worn.

As a Hoe, Cultivator, Rake and Plow.—With the Seeder removed it forms identically the same implement as the Double Wheel Hoe, etc., described above, and it is supplied with the same outfit. **Price** complete, as shown in the cut above, **$9.50.**

WEEDER ATTACHMENT

FOR THE "IRON AGE"

Double Wheel Hoe, etc., or The Combined Drill.

Especially valuable for destroying young weeds just after a rain, or for following the flat hoes to pulverize the soil. **Price, $1.00** per pair.

LAND-SIDE PLOW ATTACHMENT

FOR THE "IRON AGE"

Double Wheel Hoe, etc., or The Combined Drill.

This attaches as shown in the cut below. making a perfect hand-plow; throws a strong furrow. **Price,** plow and connection, **80c.**

The "GEM OF THE GARDEN" Single Wheel Hoe and Cultivator.

Price plain, with five teeth only, **$3.00**; or complete, with Scuffle Hoes, Plows and Cultivator Teeth, as shown in cut, **$4.25.**

A WONDERFULLY POPULAR GARDEN IMPLEMENT.

No greater proof can be given of the popularity of a tool, or of its intrinsic worth, than the one simple fact of a *continued* and increasing demand for it. During the past few years we have placed about *fifteen thousand* of the "Gem of the Garden" in the hands of gardeners, and we are not aware of a single instance in which it has failed to give thorough satisfaction to the user. It is not a toy; neither is it a tool made of extremely light gray castings, calling for constant repair, the "Gem" being made up of *steel* and *malleable iron*. The set of slender stirring teeth, each stamped from one piece of steel, cannot be excelled for thorough work, especially in hard soil. The "Gem" is nicely finished, and makes a handsome implement.

EXTRAS FOR THE "GEM" WHEEL HOES.

Land-Side Plow, **75c.**

Onion Set Gatherer, **75c.**

THE "GEM" DOUBLE WHEEL HOE.

Price complete, as show in the cut, **$4.75.**

For first and second working of crops the tool is used astride the row, while in subsequent hoeings the wheels can be closed together by means of the telescopic axles, and used between the rows as a single wheel tool. With the double wheel machine we send out, as shown in cut, the side hoes for scuffle hoeing and cutting off weeds; plows for hilling, covering furrows, etc., and five narrow cultivator teeth for deeply stirring the soil. We also offer, as an attachment, a good-sized plow, with landside, for use with either style "Gem." Works splendidly for various purposes—opens a straight, deep furrow. **Price, 75c.** extra. Also, an Onion Set Gatherer, used by passing under the row. Makes a very good scuffle hoe. Can be used with either style "Gem," though more particularly adapted for the Double Wheel. **Price, 75c.** extra.

The UNIVERSAL TWO-ROW ONION DRILL

This is the machine so largely used in the celebrated Connecticut onion-raising districts. It sows two rows at a time, marking, sowing and covering by same operation. The two seed hoppers are on the axle of the wheels. The seed is forced out by small wheels which turn on the axle, thus making it sure to drop seeds, which it sows very accurately and evenly. The track of the outer wheel is the mark for the inside wheel to follow in returning. This is the best Onion Drill in use and will pay for itself in sowing one acre, because when adjusted it sows very accurately and evenly, thus insuring a good standing of plants in the row. No seed wasted and no hand thinning required. The wheels have 2-inch face, preventing them from sinking in soft onion ground.

No man having sown one acre to onions with this Drill would use any other. Every machine warranted to do good work and to be made with the best material and workmanship.

PRICE, No. A, drills two rows, 12 inches apart, **$8.00.**
"   No. B,   "   "   14   "   **8.50.**

THE SHERWOOD ONION PULLER.

This machine is as near perfection as any machine can be made. A curved hoe runs under the onions and a moldboard rolls them off and turns like a swivel plow at the end of the row, so as to keep the pulled ones from standing. It pulls an acre or more a day without injuring them as much as hand pulling.

Those who use it would rather have onions pulled by it than by any other method.

PRICE, $5.50.

THE UNIVERSAL ONE-BLADE HOE.

The Universal One-blade Hoe is a very useful and efficient one-blade hoe for close work in the garden. The blade is 8 inches long and adjusts itself to the work by springs. Its special use is to keep the surface of the land free from weeds. It does not stir the soil except on the surface. **PRICE, $2.00.**

THE UNIVERSAL FOUR-BLADE HOE.

Has four blades with elliptical cutting edges, fastened to arms, each working independent of the others and adjusting themselves to their work by springs. For general hoeing and cultivating it is indispensable. Will do better work with greater ease and four times as fast as a hand hoe. A light, handy hoe that has had a large sale and is very popular. **PRICE, $3.00.**

Extra blades of polished steel, 12 cents. Wing to prevent dirt from being thrown on small plants, 40 cents extra.

THE SHUART LAND GRADER

ESPECIALLY ADAPTED FOR FINE GRADING, PRODUCING A BEAUTIFULLY EVEN SURFACE, PARING DOWN KNOLLS AND EVENLY DISTRIBUTING THE EARTH IN DEPRESSIONS.

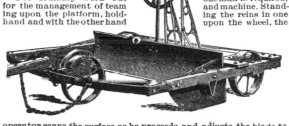

This machine is borne on three wheels, the front one being a caster wheel, allowing the machine to be turned short about. The blade is raised and lowered by a hand-wheel and can be locked at any point by a spring bolt operated by the foot.

The blade is sharpened on both edges and is reversible. A pole holds the machine off the horses, the bolt attaches to the king-bolt of the caster frame, is independent of the draft-chain and subjects the team to no strain whatever in turning.

For ordinary work but one man is required, for the management of team and machine. Standing upon the platform, holding the reins in one hand and with the other hand upon the wheel, the operator scans the surface as he proceeds and adjusts the blade to the cut required. The scraper filled, he carries the load to its destination, and without stopping the team, readjusts the blade for spreading the earth, which passes beneath the blade in a smooth sheet as the machine proceeds. The earth can be dropped in a heap or can be spread as desired.

The blade is 5½ feet long by 14 inches in width, and as the fenders prevent the earth from escaping at the ends of the blade, the capacity of this machine for moving earth is much greater than that of other scrapers of equal blade area. The lightness of draft in view of the amount of earth moved is accounted for by the smooth, even cut with which it gathers its load and in the fact that the load is in a measure borne by the wheels.

PRICE, No. 1, $48.00.
"   No. 2,   53.00.**}   Whiffletrees, extra, **$3.00.**

THE SHELDON WHEEL HOE

A low-priced but very efficient little implement for flat or scuffle hoeing. Either in gardens, where it cuts off all weeds, or for cleaning walks of weeds and grass it is equally useful. It is a very easy-working tool with a sharp steel blade, a 7-foot handle and wooden wheels. Total weight only 4 lbs.

PRICE, $1.50 EACH.

THE WEED SLAYER.

A very light and handy wheel weeding hoe, with sharp, steel, flat cutting blade that will cut off all weeds, grass, etc., below the surface. It can be run from ¼ inch deep to 1½ inches, as desired, and cuts 7 inches wide. Handles adjustable in height to suit tall or short person.

PRICE, $2.00 EACH.

SURFACE GRADER.

The steel blade measures 30 inches long by 15 inches wide and is used for removing the plowed ground from the sides of the road. Weight, 60 lbs.

PRICE, $8.00.

SURFACE GRADER.

ROAD LEVELER.

ROAD LEVELER.

The **Road Leveler** has a steel blade ¼ inch thick by 4 x 72 inches. Stamped steel seat.

For smoothing rough roads of any kind. In the spring, when the frost is first out of the ground, by merely driving once or twice over the roads the ridges are cut down, the ruts filled up, and the roadbed put in temporary good order. It will pay for its cost in one day's use. Weight, 150 lbs.

PRICE, $12.00.

SOLID STEEL SCRAPERS, WITH RUNNERS.

These scrapers are made of a *single sheet of steel*, pressed into the best and most practicable shape for working.

The runners on bottom can readily be replaced when worn.

No. 1, width 32 in.; capacity 7 ft., $9.50.
"   2,   "   29 in.;   "   5 ft.,   9.00.
"   3,   "   26 in.;   "   3 ft.,   8.50.

SOLID STEEL SCRAPER, WITH RUNNERS.

The Genuine (ADRIANCE, PLATT & CO.'S) Buckeye Mowers.

THESE famous and reliable machines are the standards of America—strictly high class in every respect. Made scientifically correct, embodying all practical improvements to date, while the older principles of proven superiority have been retained. The frame is symmetrical, light, simple but thoroughly strong, and is fitted with roller bearings and brass composition bushings. The driving wheels are large and wide apart, clearing the swath and carrying all weight, well divided to give equal power to either wheel. The driving wheels do not lift when the bar strikes an obstruction, insuring safety. The self-adapting finger bar is carried, not dragged. There is no side draft, and smooth cutting is assured. The height of cut is quickly adjusted, and can also be varied while mowing by the tilting lever—the levers are easily operated by hand or foot. It is the lightest draft mower on the market.

Prices Two-horse Buckeye Mowers.
4½-foot cut..$43.00 | 5-foot cut...$44.00

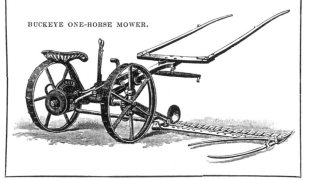

BUCKEYE ONE-HORSE MOWER.

Buckeye One-horse Mower.

THE BEST AND ..
.. MOST POPULAR
.. One-horse Mower made.

This mower is made entirely from patterns especially designed for a small machine, and is so proportioned as to secure the necessary strength consistent with lightness of draft. It has all improvements —hand and foot levers working independently or together—and is a very popular and satisfactory machine. Is especially useful on rough or obstructed ground, where a team and a long cutter bar is troublesome to manage; and is also useful on large lawns, leaving a clean, neat stubble, even in short grass.

Price One-horse Buckeye Mower.
3½-foot cut....................................$40.00

The Genuine (ADRIANCE, PLATT & CO.'S) Binders and Reapers.

These important implements are the highest types on the market to-day—thoroughly good in every respect—embodying all of the good points of older machines, with the practical improvements to date.

The Adriance Binder.

The most compact, lightest and handiest binder made, and better adapted to the wants of farmers than any other type. It is simple, light, well-proportioned, well-balanced. Working parts move slowly; limited elevation of grain; novel rear discharge of bundles; perfect separation; gentle handling of grain; clean delivery; great capacity and range of work; adapted for side hills, economy of power and time; light draft, durable and portable. It has a high reputation among farmers, and has won numerous awards both in this country and in Europe. **Price, 5-foot cut........$118.00**

THE ADRIANCE BINDER.

The Adriance Reaper.

THE BEST OF ALL REAPERS.

Perfectly balanced. Simple, strong, of light draft, easily operated, and is the most serviceable and reliable for satisfactory work in all crops in any condition. It is largely used as an ensilage harvester, for which it is better adapted than any other reaper. It embodies all of the good points of the older machines, with all of the latest improvements to date.
Price, 5-foot cut...............$68.00

BINDER TWINE.

Pure White Sisal, made from long fibre, smooth and uniform. 500 ft. to lb. In 60-lb. bales, at 9c. lb.
Strictly Pure Manilla. 650 feet to lb. In 60-lb. bales, at 10½c. lb.

THE ADRIANCE REAPER

BULLARDS' IMPROVED HAY TEDDER.
THE BEST TEDDER ON THE MARKET.

For turning and spreading hay this machine has no superior. It tosses up the grass lightly, leaving it crossed in every direction, so that it can be quickly and evenly dried. It will thoroughly turn and spread 4 acres an hour, and by doing it so rapidly the hay can be turned several times in one day, allowing it to be cut, cured and stored *in one day*; and besides it improves the quality of the hay very much. Bullards' Hay Tedder is the best machine for the purpose on the market. **PRICES.**

6-fork Spreader with shafts for 1 horse......................$35.00
8- " " pole for 2 horses................................... 42.00

SYMM'S PATENT HAY AND GRAIN CAPS.

Pronounced by the Hatch Experiment Station to be the best Cap of the several tried by them.

Thoroughly waterproof wood fibre Caps, light, efficient and thoroughly practical, and will last a lifetime. They nest together for carriage and storing, are 4 feet in diameter and 18 inches deep. By cutting hay in bloom, letting it wilt and then cocking it and covering it with these Caps, which keep off rains and dew, protect from hot sun during the day, and retaining the heat in the cock at night, a superior quality of hay can be made. These Caps are not blown off by the wind—do not require fastening down. One hundred of them, nested, take up a space 3½ feet high, and will cover 6 to 8 tons of hay. These Caps are also valuable for covering potatoes, roots and apples, allowing them to sweat before storing or barreling.

Price, 60c. each, $6.00 doz.

PATENT____
SICKLE GRINDER.

IT GRINDS BOTH SIDES OF THE SECTION AT ONE OPERATION

with both bevels the same as a new section. It grinds the heel as perfectly as the point, and will last a lifetime. We use corundum stone, almost as hard as a diamond, and it grinds equally well, wet or dry.

It can be used for grinding other edged tools, such as axes, knives, sickles, etc.

Price, $5.50.

Improved Sulky Hay Rake.

A very important feature in a Rake is the way it gathers the hay. Some roll it over and over, twisting up into a rope. With this improved rake of ours the hay is slid along the ground, and, when dumped, the damp under portion is left on top. The shape of the teeth is such that they do not scratch the ground, but rake the stubble and leave all rubbish on the ground. It has a most accurate and most durable dumping arrangement, positive at all times and under perfect control. The driver simply presses the dump lever lightly with one foot, and the teeth rise high in the air and return at the proper time without any effort on the part of the driver. These Rakes are fitted with oscillating steel

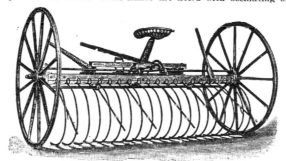

cleaner rods; it is dumped by either wheel alone, and can be dumped in any position the Rake may be drawn; large, easy steel seat that can be raised or lowered to suit driver; a foot lever by which the teeth can be held down to their work in carrying an extra large load, or the same lever will hold the teeth up for driving on roads and across fields; no springs to break—strong; and simple enough for a boy to manage.

PRICES.

8-foot Rake, with shafts, 20 teeth, $24.00 ; 24 teeth, $25.00
9- " with pole or shafts, 22 " 26.00 ; 26 " 27.00
10- " with pole, 24 " 28.00

Improved Wooden Revolving Horse Hay Rake.

For raking heavy grass the Revolving Rake has no superior. For this particular purpose we recommend it as being even better than the Wheel Rake. They are of the most approved kind, and are made in the most substantial and durable manner. 16 hickory teeth. We pack them snugly for shipment. **Price, $5.50.**

Hay and Grain Gleaning Rake.

Has a broad light head, 3 feet long, with upright teeth to prevent hay, etc., from going over back. Very useful for gleaning hay and grain fields.

Price, Each, $1.25.

THE THOMAS HAY LOADER.

This Loader enables the farmer to put up the greatest amount of hay in the least time and with less help, and permits the making of best quality hay. It will put on a load in 5 minutes, if desired. For simplicity and lightness of draft it has no equal. It rakes 8 feet and narrows to 6½ feet at top. This Loader takes hay direct from the swath; it is not necessary to rake the hay into windrows, though it will rake and load from light windrows. An adjustable fork protector prevents the forks from digging into the ground and keeps them at an even distance from the ground; they can be lowered or raised into any desired position. It is well made, strongly braced and durable. Hay Loaders are only adapted to level lands. **Price, $55.00.**

PORTABLE DOUBLE-ACTING BALING PRESS.

The best Baler for hay and straw on the market—simple, strong, durable and rapid. The machine is 15 feet long, and a lever 14 feet long is attached to the end, to which a horse or light team is hitched. They travel less than a half circle, then return; a charge of hay is forked in at each turn of the horse. It makes bales 14 x 18 inches and of variable lengths, as desired, is continuous in action, the bales being discharged as soon as made, and tied through the open bale chamber without stopping the horse. It bales from 10 to 14 tons of hay a day, tucks in all loose ends with the automatic folder, and makes a bale solid, smooth and neat on all sides.

Price (f. o. b. factory), **$280.00.**

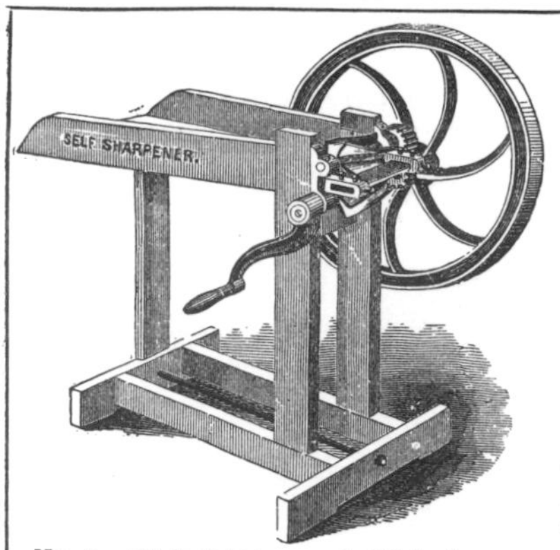

SELF-SHARPENING FEED CUTTERS.

These old reliable machines still retain their wonderful popularity; they cut easier and faster than any other of like price, using same power, and cutting as short as this. The large sizes will cut faster by hand than any other cutter, without regard to price. They are easier sharpened and repaired than any other Self-Feeding Feed Cutter; will generally cut well from three to five years without grinding; a clean, uniform cut, and do not clog.

All sizes are good for hay and straw, and while the smaller machine will cut cornstalks, yet for this purpose nothing smaller than No. 3 should be ordered.

No. 0.—6½ inch knives, cuts 1¾ in. long $9.00
" 10.—6½ " " " 1 " " 10.00
" 1.—7 " " " 1⅝ " " 11.00
" 2.—7½ " " " 1⅜ " " 11.50
" 3.—7½ " " " 1¾ " " 13.50

THE ROSS (FRONT CUT) ....FEED CUTTERS.

These machines have a high reputation. While the first cost is more than for others of same capacity, yet purchasers will find them the best investment. They are powerful, very durable, cut very rapidly, turn easily, and will cut feed in various lengths from hay to coarse corn fodder.

The oscillating rollers insure an even and perfect feed; the shear cutters are of the finest steel, easily sharpened and quickly adjusted.

PRICES:

No.	Power.	Lengths of cut. In inches.	Capacity 1 in. cut. per min.	Price
06	Hand Power	¼, ½ and 1	1 bushel	$20.00
8	" "	⅛, ¼, ½ " 1	1½ "	30.00
10	Combined Hand and Power	⅛, ¼, ½ " 1	2 bus. by hand	35.00

The last two sizes are fine poultry feed cutters.

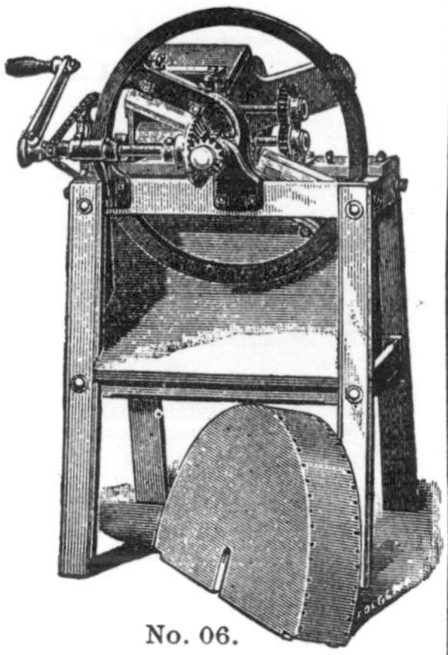

No. 06.
ROSS FRONT CUT FEED CUTTER.

THE "CHEAP" HAND LEVER CUTTER.

This is a strong framed machine, and the most simple cutter made, and we commend it to all who have but a moderate amount of feed to cut. It will cut 15 bushels an hour. We furnish all with the gauge-plate, with which short feed can be cut much faster than without it. It can also be changed to make any length of cut. Price, $3.50.

THE "CHEAP" LEVER CUTTER.

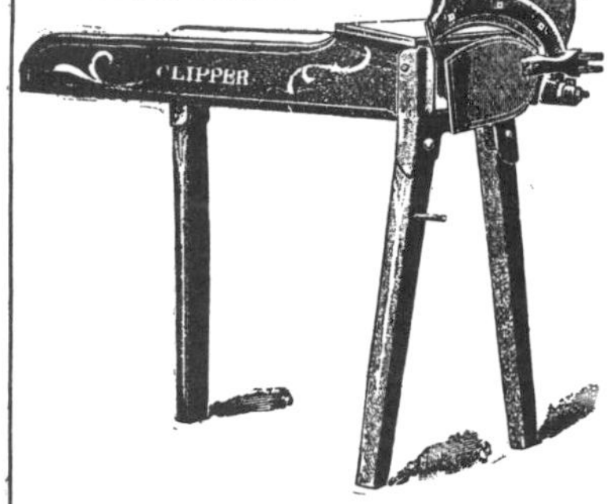

THE HIGH GRADE LEVER CUTTER.

This is a strong braced frame machine made in a durable manner, with a 16½-in. hawk's-bill knife of best cutlery steel. The machine has large capacity and is especially recommended for cornstalks, as well as straw, hay, etc. Price, with gauge-plate to regulate the length of cut, $10.00.

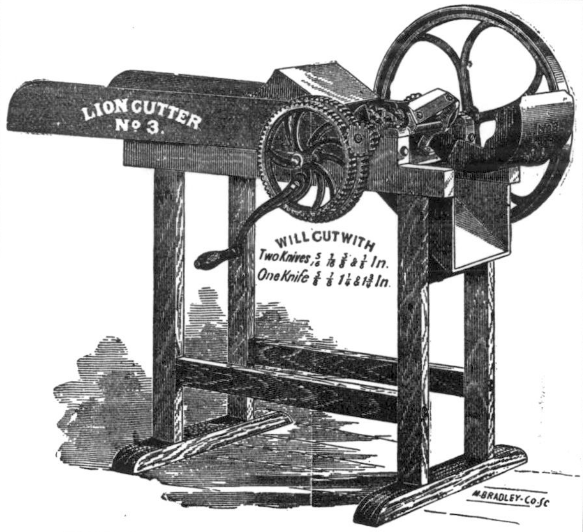

THE LION FEED CUTTER.

A self-feeding Cylinder Cutter, which will make different lengths of cut, an advantage greatly appreciated by users of Feed Cutters. The machines are made with two revolving knives, and by a simple device they can be changed in less than a minute so as to cut a different length. The feeding apparatus adapts itself to large and small bodies of feed. For the length of cut these are the most rapid Cutters made.

The following are all hand machines; power machines will be quoted on application.

No. 1.—6¼ inch knives, cuts ½, ¾ and 1⅛ inches $16.00
" 2.—7¼ " " " ½, ¾ and 1⅛ " 19.00
" 3.—8½ " " " 5/16, 7/16, ⅝ and ⅞ inches 25.00
" 4.—10 " " " 5/16, 7/16, ⅝ and ⅞ " 30.00

The lengths of cut given above may be doubled by removing one of the detachable knives.

FODDER SHREDDER.

For Making Cornstalk Hay.

A new and superior method and vastly better than the usual way of cutting. This machine has a set of saws (instead of knives). When the fodder has passed the saws it is left in the best possible condition, as it is torn into irregular fragments, so fine that there is no waste

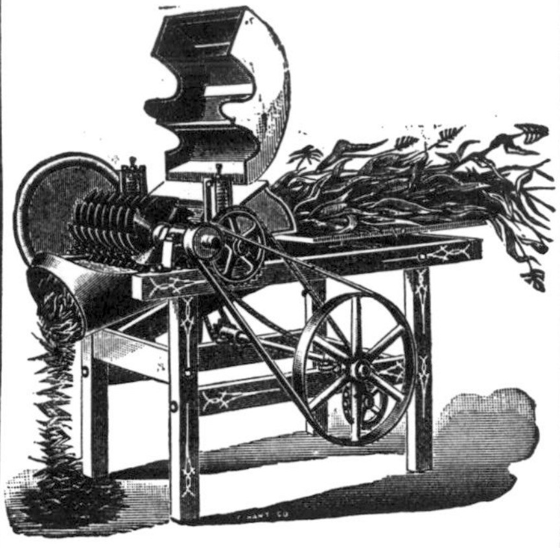

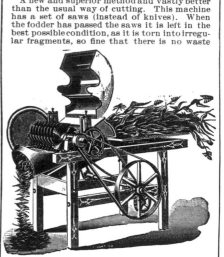

when fed out. So soft or spongy is the mass that it will absorb a large quantity of water, and by mixing with shorts or meal makes a complete food for stock.

These machines are built with Snapping Rolls or large Feed Rolls, as preferred; with Snapping Rolls, machine will completely husk 50 to 75 per cent. of the ears.

No. 1A.—6 saws, 12 in. cut, power 2 to 3 horse, $80.00.

Swinging Carrier, 12 ft. long $36.00
Ear Carrier 22.00

(*Larger sizes quoted on application.*)

ROSS LITTLE GIANT POWER CUTTERS AND CARRIERS.

For Hay, Straw, Stalks and Ensilage.

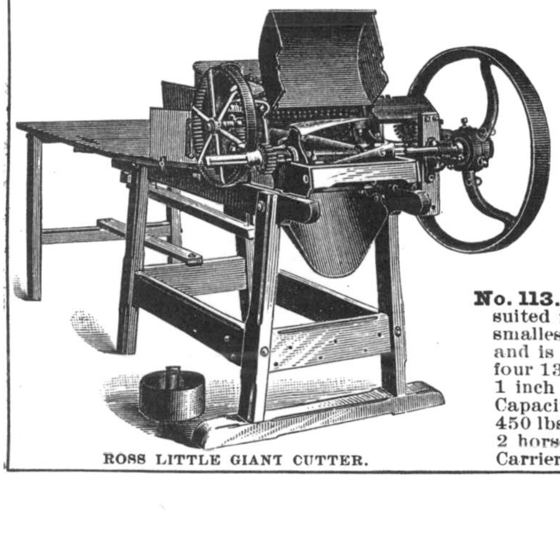

ROSS LITTLE GIANT CUTTER.

These improved machines have all of the latest devices, and are acknowledged the best and foremost Cutters made.

No. 9 has two 9 inch knives, 9 inch throat, capacity 1,200 lbs. dry fodder per hour. It is provided with crank for hand use, though 1-horse power is recommended. Weight, 250 lbs. Price, $40.00, or with 2 balance wheels and pulley, $45.00. 12 ft. Swivel Carrier, extra, $30.00.

No. 111 has four 11 inch knives and 11 inch throat; cuts ¼, ½ and 1 inch with the 4 knives and 2 inches with 2 knives; handle for hand use goes with each machine, though for much use it requires 2 horse sweep or tread power. Capacity, 2,500 lbs. dry fodder per hour. Weight, 375 lbs. Price, $50.00, or with two balance wheels and pulley, $55.00. 12 ft. Swivel Carrier, extra, $36.00.

No. 113. This is the best size for general work, being suited for either small or large quantities, and is the smallest machine adapted for regular ensilage cutting, and is supplied with the regular ensilage table; has four 13 inch knives and 13 inch throat; cuts ¼, ½ and 1 inch with the 4 knives or 2 inches with 2 knives. Capacity, 3,000 lbs. dry fodder per hour. Weight, 450 lbs. Power required, 2 to 4 horse sweep, or 1 to 2 horse tread power. Price, $65.00. 12 ft. Swivel Carrier, extra, $42.00.

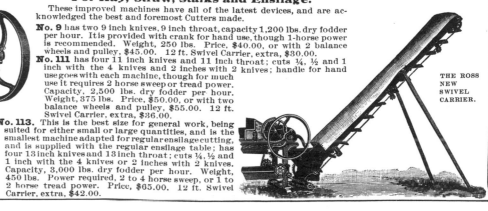

THE ROSS NEW SWIVEL CARRIER.

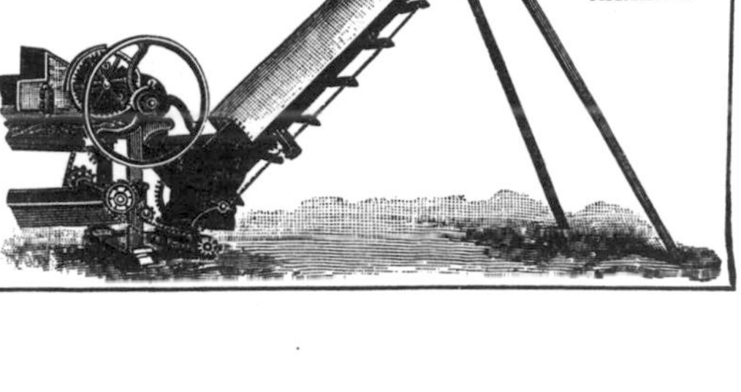

GEM CLOVER CUTTER, WITH SHORT LEGS.

TROWBRIDGE CLOVER CUTTER

GEM CLOVER CUTTER, ON LEGS.

The Gem Clover Cutters.

All iron and steel. Cuts clover or other poultry fodder, either green or dry. Cuts clean, fine and easy. Has a screw feed, regulating the length of cut. Knives removable for sharpening. Prices, with short legs to fasten on a bench or table, $9.00; with long legs, $10.00.

CLOVER AND POULTRY FOOD CUTTERS.

Progressive poultry specialists now consider it an absolute necessity to supply chickens, etc., with green cut food, not only during the winter months but also in the summer, where flocks are confined and are not able to procure a sufficient quantity themselves. We herewith offer a variety of machines for the purpose adapted to various sized flocks.

The Trowbridge Lever Clover Cutter.

For cutting clover (green or dry) for poultry; also for cutting celery, cornstalks, cabbage, etc. Can be gauged to cut from ¼ inch up. Cuts back and forth; two cuts to a complete motion. Cuts enough for fifty fowls in one minute. Price, $3.00.

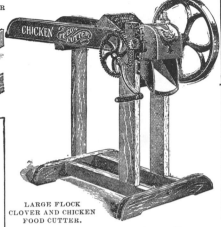

LARGE FLOCK CLOVER AND CHICKEN FOOD CUTTER.

Large Flock Clover and Chicken Food Cutter.

Invaluable for extensive poultry raisers. This is a highly efficient machine for cutting clover, hay, celery or other feed into extremely short lengths. It cuts ¼, ⁷⁄₁₆ and ¾ ins. It is made with three knives 6¼ ins. long, and a self-feeding comb roller which straightens the feed as it is presented to the knives, making a more uniform length of cut. It cuts very rapidly, being one of the fastest machines of its size now upon the market. Price, $16.00.

     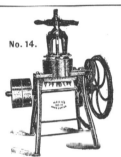

No. 6. No. 2. No. 8. No. 10. No. 14.

SIMPLEX LEVER BONE CUTTER.

POST BONE CUTTER.

STANDARD BONE CUTTER

DOUBLE HAND BONE CUTTER.

SMALL POWER BONE CUTTER.

LARGE POWER BONE CUTTER.

MANN'S "SMALL FLOCK" BONE CUTTERS.

No. 1 B.

No. 1 B. M.

No. 4 B.M.

Mann's Superior Bone Cutters

WILL CUT, FINE or COARSE, EITHER GREEN or DRY BONES.
Various Styles, for Poultry Flocks of All Sizes.

These Bone Cutters have attained a reputation for durability, ease of operation, rapidity of work, and adaptability for cutting bone in all conditions, that place them easily in the lead of all other bone mills. Almost any mill will cut or grind dry bone satisfactorily, but it is the fresh green bone that makes chickens grow and hens lay, and these machines are made particularly for cutting such. They will cut fresh bones with meat and gristle on them; dry bones; pressed scraps; corn or corn on the cob; potatoes, turnips and vegetables of all kinds, and all without clogging. These machines are cutters, not grinders; they may be operated by boy or man; they run easy and cut fast; have an automatic feed; can be adjusted to cut fine or coarse; have the finest quality steel knives, that can be removed and sharpened by any one, and many other points of merit, and are warranted to do everything claimed for them.

		Capacity, Lbs. an Hour.	Weight, Lbs.	P. H. & Co.'s Net Price.	Mfr.'s Price.
Small Flock Cutters	No. 1 C, with Crank Handle	10	30	**$5.00**	$6.25
	" 1 B, " Balance Wheel	15	55	7.00	8.75
	" 1 B.M.,on Iron Stand, with Balance Wheel,	15–20	80	10.00	12.50
Popular Cutters	" 4 B, with Balance Wheel	25–35	80	13.00	17.50
	" 4 B.M.,on Iron Stand,with Balance Wheel,	25–35	105	16.00	21.25
Post Bone Cutter,	" 6, can be fastened to any post	30–60	100	16.00	20.00
"Standard" Cutter,	No. 2	30–60	140	18.40	23.00
Double Hand Cutter,	" 8	45–75	180	22.40	28.00
Small Power Cutter,	" 10, ½ to 1 Horse Power	60–100	175	26.00	32.50
Large Power Cutter,	" 14, 1½ to 4 "	150–250	425	76.00	96.00
Lever Handle Cutter,	Simplex		40	10.00	12.50
" " "	Simplex M., mounted on Iron Stand		70	12.00	15.00

No. 1 C.

Mann's "Popular" Bone Cutter
The New No. 4 Machine.

For medium to good-sized flocks this is the best size. It surpasses all for ease and rapidity of cutting. Cuts finer than any other. Fine enough for small chicks. The new automatic feed governor feeds automatically and regulates the power required according to the feed.

The "Shaker" Two-Hole Corn Sheller.

FOR EITHER HAND OR POWER.

This is the most complete, convenient and easiest-running Sheller of equal capacity on the market. By hand it will shell from 200 to 250 bushels per day, and by power from 400 to 500 bushels. The balance wheel and a 20-inch pulley wheel are combined, so that it can be used either way at any time.

The Sheller is provided with a shaker for separating the corn from cobs and impurities, and cleans the corn in the most thorough manner.

PRICES.

Plain (*as shown in cut to left*), **$15.00**, or with **Feed Table, $16.50.**

With Elevator, Double Discharge Bagger and Feed Table (*as shown in cut to right*), **$30.00.**

THE "SHAKER" CORN SHELLER.

THE "SHAKER" TWO-HOLE CORN SHELLER WITH FEED TABLE, ELEVATOR AND BAGGER.

The "Black Hawk" Corn Sheller.

Fastens to a box quickly and securely by clamps which are furnished free with each machine.

It is recognized as a standard sheller and no machine has ever given more universal satisfaction.

The "Black Hawk" Sheller shells corn for home use and poultry. Its construction is such that it cannot choke, is durable; malleable iron is largely used in its construction. Bearings that are subject to wear are long and thoroughly chilled. It will last a lifetime, is simple, easily adjusted and will shell clean all kinds of field corn. **PRICE, $2.25 EACH.**

The "Henderson" Corn Sheller

WITH COB SEPARATOR, FAN AND FEED TABLE.

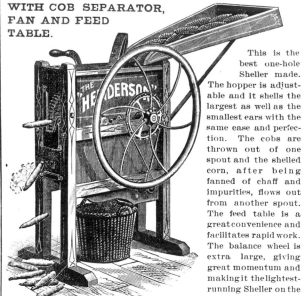

This is the best one-hole Sheller made. The hopper is adjustable and it shells the largest as well as the smallest ears with the same ease and perfection. The cobs are thrown out of one spout and the shelled corn, after being fanned of chaff and impurities, flows out from another spout. The feed table is a great convenience and facilitates rapid work. The balance wheel is extra large, giving great momentum and making it the lightest-running Sheller on the market. The gearing is all enclosed, preventing accidents. It is handsomely painted and finished.

PRICE, COMPLETE (*as shown in the cut*), **$11.25.**
"   **WITHOUT FEED TABLE,** - **10.00.**

Family Pop Corn Sheller.

This little Sheller is made of malleable iron and will last a lifetime. It may be held in the right hand and the ear of corn in the left hand. The shelling is done easily and without any scattering of the corn.

All lovers of pop corn who have had sore fingers from shelling the corn from the cobs should send for one of these Family Pop Corn Shellers.

**PRICE, 25 CENTS,
OR MAILED FOR 30 CENTS.**

The Burrall Improved Corn Sheller.

An all-iron and very durable one-hole Sheller. It shells both large and small ears and separates the corn from cob perfectly. One man can turn and feed at the same time. Capacity, 100 bushels per day. It has been a popular sheller for years and with the present improvements is better than ever.

PRICE, $7.00.

The Improved Clinton Corn Sheller.

A one-hole sheller; the cheapest satisfactory Sheller. It will shell 10 bushels of shelled corn per hour, but does not separate corn from cob. It is well known, having been on the market for over half a century. It is well suited for all ordinary grades of corn, but for extra large eared corns the "Henderson" or "Shaker" will do better work.

**PRICE, CLINTON SHELLER,
With 1 Balance Wheel, $4.50.**
"   "   2   "   "   5.00.

The "ENTERPRISE" GRINDING MILL.

A good general mill for farmers, poultrymen, etc., and for strength and durability it is unexcelled; the grinders are warranted equal to steel. Capacity, 1½ bushels of corn per hour. It grinds corn, grain, roots, bark, shells, dry bones, chicken feed, salt, etc. It is not intended for grinding green bones, for these can only be shaved or cut, but not ground. The dimensions of the mill are as follows: Height, 17 inches; width, 8½ inches; balance wheel, 19 inches diameter; throat, 3½ inches diameter; weight, 60 lbs. **Price** (No. 750), as shown in the cut, **$7.00.** No. 650, same as the above, excepting that it is made to screw to a wall or post, **$7.00.**

FARMER'S CORN and COB MILL.

Crushes and grinds ear corn or shelled corn.

Will grind corn and cob, shelled corn, oats, rye, corn and cob with husk, etc. Can be used with either one or two horses. Grinds from 3 bushels fine to 12 bushels coarse per hour. The grinders are instantly changed by turning a small thumb-nut, to grind fine or coarse. The grinding plates are cast steel, are very durable and will last for years, and can be replaced at a cost of $4.00. **Price, on iron legs, $25.00.**

FARM MILL, No. 3,

...FOR

GRAIN AND BONE.

For Two or Four Horse Power.

This is a most complete mill for the farmer and poultryman. It will grind shelled corn, corn on the ear, and all small grain. It will also grind bones, raw and greasy or dry. The burrs are so constructed that it is not necessary to run bones through the mill more than once; grind it fine the first time. The burrs are held in place by a weight and lever. Capacity, 60 to 100 lbs. of bone an hour; grain, 5 to 12 bushels an hour. Weight, 325 lbs.

Price, - $60.00.

Patent Bag Holder

AND FUNNEL COMBINED.

It consists of a metal funnel on which are four hooks which hold the bag open. The funnel can be raised or lowered on the standard for bags of any length. It may be placed on platform scales for bagging a certain weight of grain.

Price, - - $3.00.

Patent Bag Holder.

The "BANNER" ROOT and VEGETABLE CUTTER.

The "Banner" Root Cutter is, without question, the best ever invented; it contains features found in no other machine; it has a self-feeder; *separates the dirt from the cut feed,* and leaves the cut food in such condition that stock cannot choke. The pieces are cut in long, half-round slices, and are not crushed, ground or torn; even young lambs can safely be fed with roots cut with the "Banner" cutter.

PRICES:

No. 20.	Hand power; capacity, 30 to 50 bu. an hour;	$11.00
" 15.	The above fitted with pulley for power; capacity, 60 to 80 bu. an hour, - - -	13.00
" 16.	Power Cutter; capacity, 120 to 180 bu. an hour,	28.00

FARM GRIST MILL,

...FOR...

One or Two Horse Power

Though it can be turned by hand.

It grinds corn on cob, fine corn meal, and Graham flour for table use. Capacity, 5 to 8 bushels an hour. It runs very easy and grinds fast and fine, and has an automatic feed. It is a very durable mill, the burrs being made of the hardest material, and can be replaced for $1.00 per set, though an extra set is furnished free with each machine. Weight, 150 lbs.

Price, - $25.00.

Bag Trucks.

Made from best oak or ash timber with bent handles.

Price, No. 2, - - **$3.75.**

"CHAMPION" ROOT CUTTER.

It is strong and durable, very simple in its construction, and very rapid. The hooked teeth, in revolving, pass between stationary knives and catch hold and tear the roots, etc., into small pieces. The hopper will hold about a bushel of turnips, etc., which can be cut in one minute. Pumpkins, turnips, beets, carrots, etc., can be cut with this machine, and fed to horses, cattle, sheep and calves, without danger of choking them.

Price, - - $6.00.

THE JOHNSON FAN MILL.

This is acknowledged the best farmer's fan mill on the market. It runs light, yet is strong and durable. It runs the sound, clean grain to one side of the mill, while small seeds and sand pass out of the other side, and the heavy chaff and heads are caught in the receptacle behind. It will not only clean grain, but grass seeds as well, and all in the best manner.

Price, - - $15.00.

THE HUTCHINSON ... ❧ FAMILY CIDER .. ❧ AND WINE MILL.

For those who only make limited quantities of cider or wine, this old favorite machine is unexcelled. It will grind from *eight to ten bushels of apples*, and from *ten to twelve bushels of grapes, currants, etc.*, per hour. The curb (or press) will contain the pomace of one bushel or more of apples. One man can make with it from *two to three barrels of cider*, or from *one hundred to one hundred and fifty gallons of wine per day*, while it is always ready to make a pitcher or bucket of cider in a few minutes. Weight, about 150 pounds.

PRICES.

With 11 x 11-inch curb.............................$9.00
"   12 x 12   "   " 10.00

CIDER AND WINE PRESSES ❧ ❧ ❧

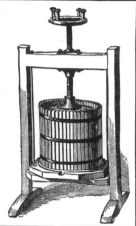

These new style presses are much stronger and more durable than those usually sold. The upright side posts are made wide at the bottom, giving a bearing the full width for the bed timber. They also have two rods on each side instead of only one, as usually used, thereby greatly adding to the strength of the frame. The screws are made of steel.

PRICES.

	Size of Tub.	Capacity, about	
No. 1.	10 x 9 inches.	2 gallons.	$5.00
" 2.	12 x 11 "	5 "	6.00
" 3.	15 x 13 "	8 "	10.00
" 3½.	....	12 "	20.00

BARREL-HEADING PRESSES. ❧ ❧ ❧

"Screw"
Barrel-head
Press.
$1.50 each.

"Lever"
Barrel-head
Press.
$1 50.

THE NEW DEPARTURE ❧ CIDER MILL. ❧ ❧

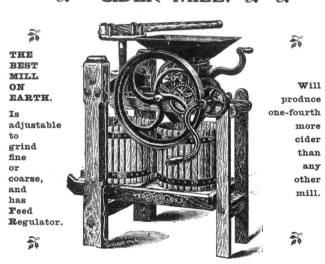

THE BEST MILL ON EARTH.

Is adjustable to grind fine or coarse, and has Feed Regulator.

Will produce one-fourth more cider than any other mill.

This cider mill has established a reputation over all others, and has proved itself to be the leading mill. It has been awarded the **First Premium** over all competitors at state and county fairs, where it has met mills of every character, and beaten all. It is constructed on an entirely different principle from other mills, which either *grate* or *cut* the apples, leaving the larger portion of the pomace in lumps, from which the juice cannot be extracted by the press. In this mill the two lower rollers are cast with alternate ribs and grooves interlocking, to draw in the apples, and the fruit is mashed between the smooth segments, *breaking thoroughly all the cells*, so that the cider is entirely extracted in the press. It will produce one-fourth more cider than any other mill from the same quantity of apples. For light running and fast grinding it has no equal. **Price, $22.00.**

COOK-STOVE ❧❧ FRUIT-DRIER OR EVAPORATOR. ❧

This is the latest, cheapest and best machine of its kind on the market. It can be used on any kind of stove. Its dimensions are: base, 22 x 16 inches; height, 26 inches. Containing eight galvanized wire-cloth trays. Weight, 25 lbs. No extra fire is required. It is easily set off and on the stove, as needed, whether empty or filled with fruit. With it you can evaporate all kinds of fruits, berries and vegetables.

Price, $6.00.

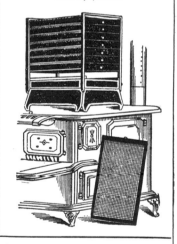

PEACH AND FRUIT SORTER.

The softest ripe peaches, apricots, plums, etc., can be graded in this machine without injury. It can be adjusted to change the size of the grades as required. It will also sort potatoes. It runs with a foot treadle, and is as easy to operate as a sewing machine. Fruit growers cannot afford to be without it. It increases the value of your fruit.

No. 1.
Capacity, 200 bushels per day.
Price, $30.00.

AMERICAN FRUIT EVAPORATORS.

The signal success of these evaporators is accounted for by reason of their simplicity, ease of management, thorough ventilation and automatic circulation. Every kind of fruit—berries, bananas, apples, pears, cherries, plums, grapes, peaches; also vegetables—corn, beans, pumpkins, etc., evaporated with these machines are of the highest quality.

No. O
EVAPORATOR.

PRICES OF FAMILY SIZES.

No. O. 6 feet long, weighs 200 lbs., capacity 3 to 5 bushels a day.....$25.00
"   1.   6   "   "   " 350 " 6 " 8 " " 50.00
"   2.   9½ "   "   " 600 " 10 " 12 " " 75.00
(*Larger sizes quoted on application.*)

One and Two Horse Sweep Power.

FOR PUMPING WATER, CHURNING, SAWING, CUTTING FEED, CORN SHELLING, Etc.

Above powers are complete, ready for use, with tumbling rod 8 feet long, and band wheel 36 inches in diameter.

Prices: 1-horse power, 11 revolutions of shaft to 1 of horse...............$35.00
  2- " " 7 " " 1 of horses...............40.00
Large 2- " " 13 " " 1 " " 50.00

Railway or Tread Powers. VARIOUS SIZES.

Small Tread Powers for Dog, Sheep, Goat or Calf. Any of the above animals soon learns to run these powers very steadily. We recommend the use of an animal weighing from 150 to 200 lbs. The steepness of the floor can be instantly varied to suit animals of different weights or to govern the speed. Price, power with brake, $15.00; power, brake and pulley wheel, $17.50.

No. 1 Power, for pony, bull or heifer, or any animal weighing from 400 to 1,000 lbs. Price, with 8-in. pulley, 5-in. face, with speed regulating governor, $65.00.

No. 2 Power, for one horse; length of tread, 7 ft. 8 in. This is strong enough to carry a large horse, and is equally adapted for light or heavy work. Price, with either 42-in. band wheel or an 8-in. diameter, 5-in. face pulley for slow speed and speed regulating governor, $100.00.

Two-Horse Tread Powers. Prices on application.

Improved "Swing Table" Sawing Machine.

The table is just balanced, swinging on centres. Simply lay the timber or wood on the table and swing it against the saw. It works quicker than "slide table" saws, and can be operated by one or two horse power.

Price, with 24-inch saw...............$38.00

SWING TABLE SAW.

"Slide Table" Sawing Machine.

For general purposes this is a better machine than the swing table, for with this boards may be ripped, as well as cutting wood for fuel, but for the latter purpose only the swing table is quicker. One or two horse power required.

Price: Slide table wood saw, with 24-in. saw,
         $38.00
 " With extra table for lumber, etc......45.00

Granite State Feed Cooker.

Cooks Food for Cattle, Hogs and Poultry. The Best Farm Boiler for Poultrymen, Stock Raisers and Dairymen.

This boiler can be used for cooking all kinds of food for hogs, cattle, horses, dogs and poultry; also for heating water when butchering hogs; for rendering lard, making soft soap and with an extra boiler of tin, for preserving fruits and vegetables, boiling cider, making apple jelly, sugaring off maple syrup and many other purposes too numerous to mention. The boiler is made of galvanized steel. The furnace is cast iron; linings sheet steel. The air between linings and outside becomes heated and passes directly under bottom of boiler. The heating capacity is thus increased and less wood is needed. Prices:

25 gallons........$15.00	40 gallons........$23.00
30 " 18.00	50 " 25.00

The "Lock Lever" Post Hole Digger.

The "lock lever" is an automatic self-lock, and locks the blade at right angles so that both hands are used in lifting the loosened earth. It works successfully in all kinds of soil where others fail. It is the easiest operated; very durable. More soil can be brought out at one time and with less labor. Scours or cleans better in mucky soil. Made of best tempered steel. Only one blade to keep sharpened. Makes a hole of any desired diameter. By the use of the compound lever operating blade ordinary stones that obstruct digging are readily loosened and removed.

Price...............$1.75 each.

"LOCK LEVER" DIGGER.

Post Hole Auger.

In soils free from large stones this tool gives great satisfaction. It is used as an ordinary auger bit, and post holes can be dug quicker with it than with any other style of digger.

Price.....................$2.00 each.

Plain Wrought=Iron Thrasher.

With Patent Shaker to Separate Grain from Straw.

Unsurpassed for thrashing wheat, oats, rye, buckwheat, rice or any kind of grain. The patent shaker separates the grain from the straw, but it does not clean out the chaff; for the latter purpose it is necessary to use with a Fan Mill. These machines are made especially to use with horse tread or other light power. They are simple, understood by any one, and quickly and easily set up and operated. The shaker can be detached in a moment, and folds up for shipment.

Prices, complete for belt, with shaker:
  22-inch thrasher,
    $58.00
  25-inch thrasher,
    $65.00
  30-inch thrasher,
    $70.00

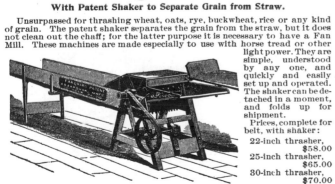

The Improved Monarch Hand Power Baling Press.

The best hand power press made for baling hay, straw, wool, cotton, etc. The double-acting lever gives enormous power with the minimum of exertion in rapidity. No steam press can pack a bale, remove it and be in position for another in less time. It can be set up by ordinary men in half an hour, and is simple, strong and durable.

PRICES:

No. 1 Hay Press makes a bale 42 ins. long, 20 ins. wide and 20 ins. deep, weighing 100 to 125 lbs. Price, $80.00.

No. 2 Hay Press makes a bale 48 ins. long, 24 ins. wide and 30 ins. deep, weighing about 200 lbs. Price, $85.00.

No. 3 Cotton Press, makes a bale 56 x 28 x 32 ins., weighing 500 to 600 lbs., $150.00.

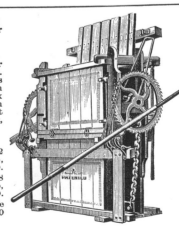

THE "STUDEBAKER" ONE-HORSE WAGON.

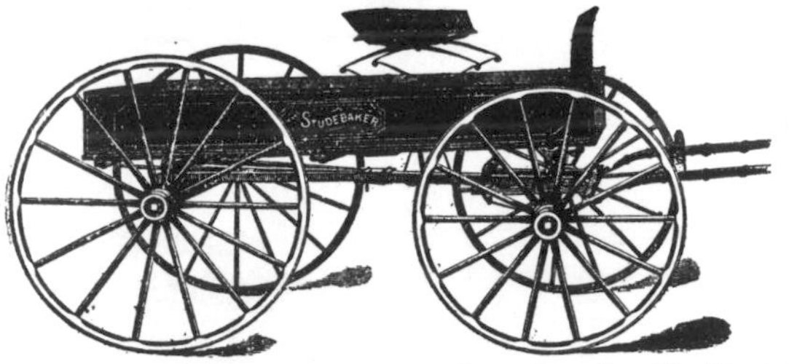

Well made of best material, and handsomely painted, varnished and striped. In ordering, please state whether wide or narrow track is wanted; wide track is 5 ft., and narrow 4 ft. 6 ins. from centre to centre of wheels on the ground.

We can furnish iron axles in place of wooden thimble skein axle on these wagons for $3.00 extra.

Sizes and Prices of Studebaker 1-horse Wagons, including Spring Seat and Shafts.

	Weight	Capacity.	Length of Box	Depth of Box	Size Axle	Prices. Narrow Tire.	Prices. Wide Tire.
Light 1 horse..	435 lbs.	1000 lbs.	7½ ft.	8 in.	2⅛ x 6½ in.	1⅛ in. **$40.00**	2 in. **$42.50**
Regular 1 " ..	560 "	1200 "	7½ "	10 "	2⅜ x 7 "	1¼ " **42.00**	2 " **45.00**

Extras: Iron Body Brake, $5.00; Bolster Springs, $5.50; Top Box, $2.50; Lazy Back for Spring Seat, 50c.

HORSE DUMP CARTS.

For Farms and Private Grounds,

And Contractors' Use.

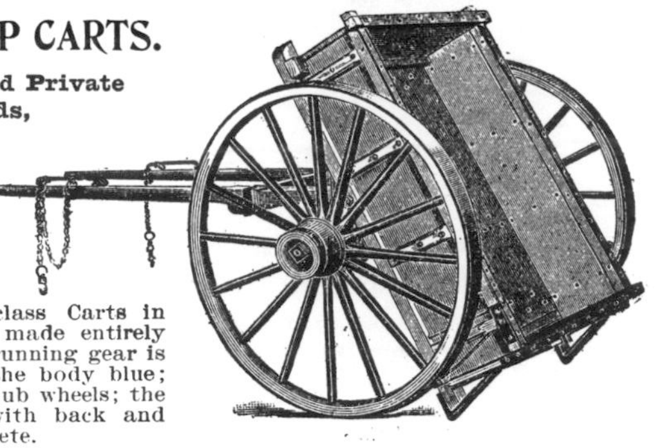

These are first-class Carts in every particular; made entirely from hard wood; running gear is painted red and the body blue; iron axles; wood hub wheels; the shafts are fitted with back and draft chains, complete.

	Weight	Capacity	Inside Dimensions, Box: Length	Width Front	Width Back	Depth	Size Axle	Height Wheel	Prices: Narrow Tire	Wide Tire
Lawn Cart ..	470 lbs.	1000 lbs.	5½ ft.	3 ft.	40 in.	10 in.	1½x 8 in.	4 ft. 4 in.	$40.00	$43.00
Farm Cart...	530 "	1300 "	5½ "	3 "	40 "	12 "	1⅝x 9 "	4 " 6 "	42.00	45.00
Contractor's	750 "	2500 "	5⅔ "	3⅙ "	3⅙ ft.	12 "	2 x11 "	4 "10 "		50.00

SPRING WAGON HARNESS, No. 49.

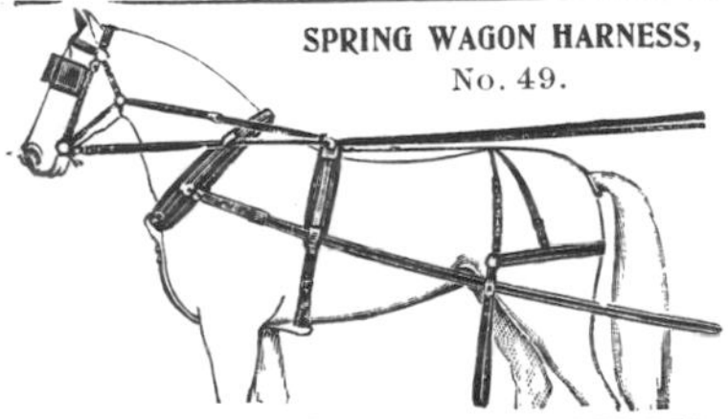

SINGLE SPRING WAGON and SURREY HARNESS, No. 49.

Bridle, with pat. leather blinds; traces, 1⅛ in.; breeching with double hip straps; round crupper; saddle 3½ in. pat. leather; lines, 1 in.; collar sheep, hog rim. **$14.00.**
Heavier grade (No. 100), $19.00.

SINGLE BUGGY HARNESS, No. 0.

Bridle, ⅝ in.; pat. leather blinds; traces, 1 in.; saddle, 2¾ in., pat. leather; lines, ¾ in.; breeching, and breast collar. **$8.00.**
Heavier grade (No. 20), $11.00.

CART HARNESS, No. 275.

Bridle, ⅞ in.; pigeon wing blinds; lines, ⅞ in.; saddle, kersey bottom; breeching, 2½ in.; collar, hogskin. **$14.50.**

CART HARNESS, No. 275.

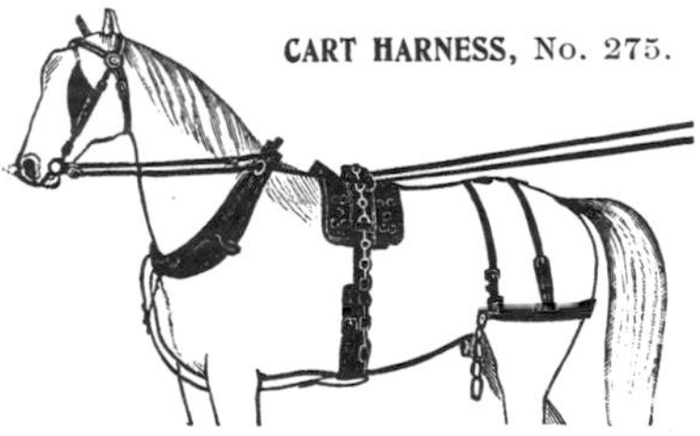

Studebaker's Standard 2-horse Farm Wagons.

These wagons are famous for their superiority, being made of best material, handsomely painted, striped and varnished. We can furnish them with either narrow or wide track. In ordering, please state which is wanted. The narrow is 4½ ft. from centre to centre of wheels on the ground, and the wide is 5 ft. Prices include whiffletrees, stay chains, neck yoke and wrench, but no seat or brake.

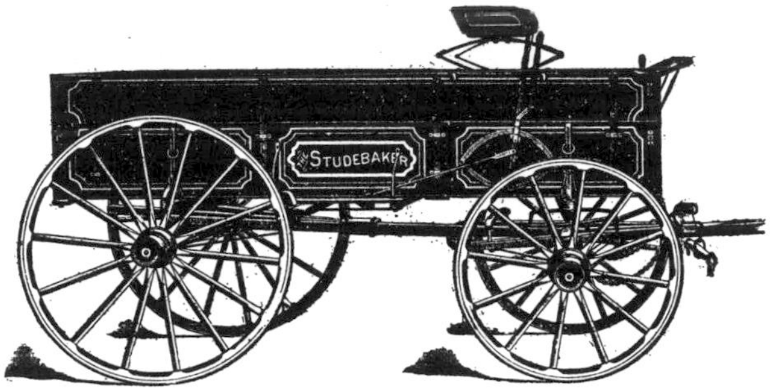

Sizes and Prices of Studebaker 2-horse Wagons.

	Weight	Capacity	Length of Box	Depth of Lower Box	Depth of Upper Box	Size Axle	Prices: Narrow Tire	Prices: Wide Tire
No. 1	800 lbs.	1500 lbs.	9½ ft.	10 in.	8 in.	2½x 8 in.	1⅜ in. **$58.00**	2½ in. **$63.00**
" 2	950 "	2000 "	10 "	12 "	8 "	2¾x 8 "	1½ " **60.00**	3 " **66.00**
" 3	1050 "	3500 "	10½ "	13 "	8 "	3 x 9 "	1½ " **63.00**	3 " **70.00**
" 4	1150 "	4000 "	10½ "	14 "	10 "	3¼x10 "	1½ " **66.00**	3 " **73.00**
" 5	1250 "	5000 "	10½ "	16 "	12 "	3½x11 "	1⅝ " **70.00**	3 " **78.00**

For the following extras add to above prices:

Spring Seat........................$3.25
" " with Lazy Back.. 3.75
Joint Brake, complete..........$3.50
Gear Brake, for gear and box, 10.00

PLOW HARNESS, No. 226.

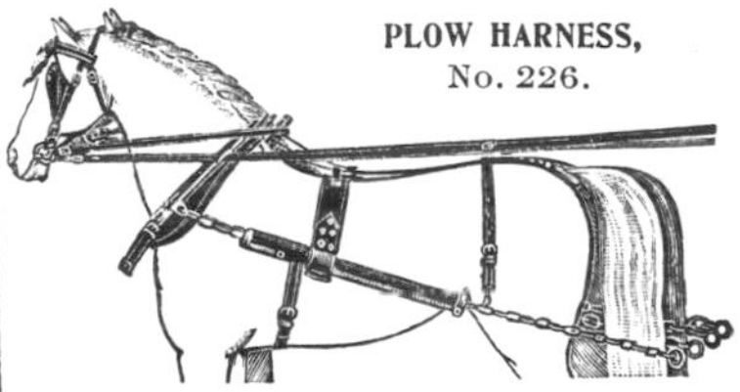

DOUBLE PLOW HARNESS, No. 226.

Traces, 7 ft. chain, with leather piping; hames, wood, with hooks; back bands, 4 in.; back straps, 1½ in. from hames to crupper; bridles, ⅞ inch, with leather blinds; collars, Scotch style, hogskin. **$22.00.**

DOUBLE FARM HARNESS, No. 233.

Traces, all leather, 1½ ins.; hames, over top wood; bridles, ⅞ in.; sensible blinds; pads, folded 1¼ in. billets; 1¼ in. back straps; 1 in. hip straps; lines, ⅞ in.; breast straps, 1½ ins.; martingales, 1½ ins.; collars, hogskin, Scotch. **$30.00;** or with breeching, $35.00.
Cheaper grade (No. 227), similar to above, excepting traces are half chain, no breeching, $24.00.

FARM HARNESS, No. 223.

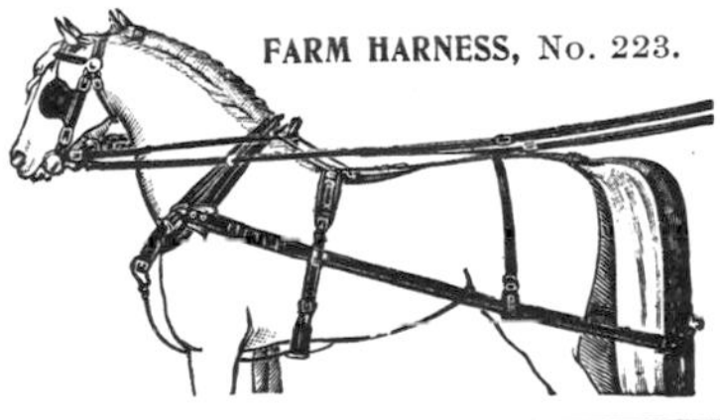

THE "LITTLE GEM" HORSE POWER SPRINKLER,

For Sprinkling Lawns, Driveways, Trees, Gardens, etc.

Will spread water 18 feet wide, or the spread may be reduced to any desired width down to one foot at the will of the operator; or it can be readily adjusted to apply one or two narrow streams at one time, directly onto vegetables or other plants in rows, thus sprinkling two rows at one time. It is the only sprinkler adapted for spreading liquid manure, as it will not clog. The valves are operated from seat, each side working independently of the other. Capacity, 150 gallons; tire, 4 ins. wide; tracks (outside), 4 ft. 10 ins.
Price, without hose, $90.00.

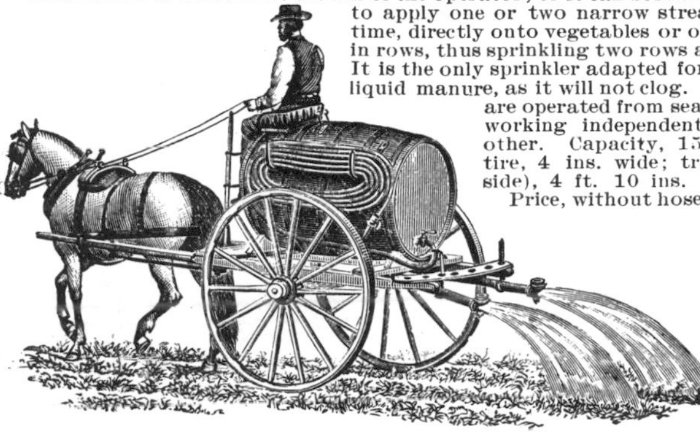

ASPINWALL PARIS GREEN SPRINKLER.

A most perfect machine for destroying Potato Bugs. It sprinkles 2 rows at a time, covering about 15 acres in a day.

The sprinkling tubes, which can be instantly shifted to the right or left by the foot, thus keeping the spray over the two rows constantly. The wheels are adjustable and can be set to suit different widths of rows. It can be used with equal facility for **sprinkling cabbage and other plants.** An even and uniform mixture of the poison is secured, and the rubber tubes that supply the liquid can be raised or lowered at will and at once thrown out of gear.
Price, $28.00.

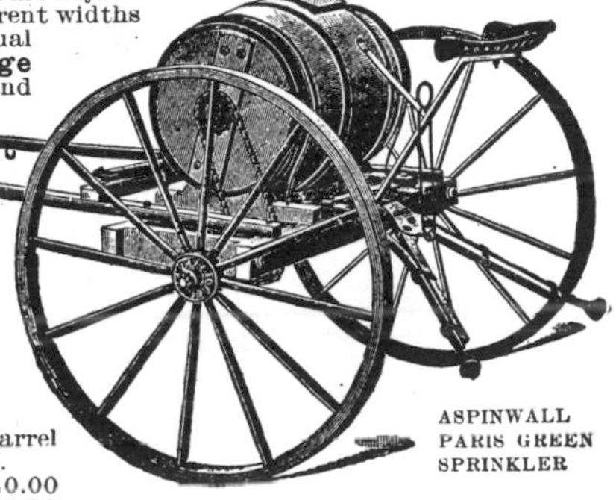

ASPINWALL PARIS GREEN SPRINKLER

Aspinwall Special Potato Sprayer.

Sprays 4 to 6 acres with 1 barrel of Bordeaux Mixture for blight.
2-row machine..........................$40.00
4-row machine.......................... 45.00

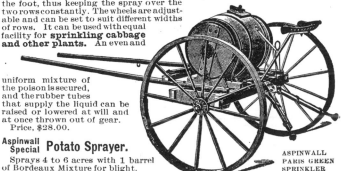

Lawn and Stable Barrow.

Extra Large
and
Extra Strong.

Our Lawn and Stable Barrow is designed for extra heavy work and has a large capacity, adapting it for manure, lawn litter, etc.; box 25 inches wide by 32 inches long by 18 inches deep.

Price, $5.00.

THE Regulation

Garden Wheelbarrow.

A superior barrow, handsomely painted and striped; iron leg braces, bolted, not screwed on, run under the legs, forming a shoe to slide on, avoiding racking the barrow; iron bands shrunk on hubs. No pine used in these barrows.

Number.	Size.	Length Box.	WIDTH OF BOX.		Box Depth.	Diam. Wheel.	Tread.	Price.
			Wheel End.	Handle End.				
No. 2.	Boy's.	21 in.	15½ in.	18 in.	9 in.	16 in.	1¼ in.	**$3.00**
No. 4.	Medium.	26½ "	18½ "	23 "	12 "	20 "	1½ "	3.50
No. 5.	Large.	28 "	20 "	24 "	12 "	22 "	1½ "	4.00

Tubular Steel Frame, Steel Tray Barrows.

The trays are made of one piece of steel the edges of which are turned over a $\frac{5}{16}$ inch steel rod, which prevents breaking at the edge. These barrows are made to dump forward, and are so constructed that at the dumping point they will not run back on the operator. The trays are about the same size as ordinary wooden canal barrows. These barrows will last a lifetime. Weight, about 70 pounds.

Price, $6.00.

Steel Garden and Farm Barrow with Tubular Steel Frame.

This style is especially adapted to wheeling dirt, manure, etc., having a deep steel tray, made in one piece, with the edges turned over a $\frac{5}{16}$ inch steel rod; the frame, of tubular steel, is so constructed that in dumping forward the barrow does not run back. Size of tray, 27x33 inches; weight of barrow, about 70 pounds.

Price, $6.50.

The "Henderson" Barrow.

This barrow is light, strong and durable, has a steel wheel and axle and oil-tempered springs, and we do not hesitate to say that it is the best barrow on the market and as well painted, striped and varnished as a buggy. While its carrying capacity is from 300 to 500 pounds, its weight is less than 40 pounds; fully warranted to stand the roughest usage.

The tire being wide makes it more desirable for lawn and garden use and a stronger wheel for the pavements. The shoe brace, running from the rear of the body to the foot of leg, is a great protection to the leg against breakage, and prevents it from sinking into soft ground. The barrow is made of selected material, and will outlast several of the cheaper, heavy, clumsy barrows. Weight, 39 pounds; body, 26 inches by 22½ inches by 12 inches deep; diameter of wheel, 20 inches; width of tire, 1¾ inches.

Price, $5.00.

Steel Tray, Wood Frame, Steel Wheel Canal Barrow.

Frame of selected hard wood; iron braces form a shoe so the barrow slides when set down, without racking it. Heavy steel wheels 17 inches in diameter; tread 1⅜ inch. Trays of one piece of sheet steel, about the size of an ordinary wood canal barrow, 28 inches long by 31 inches wide; the edges are turned over a $\frac{5}{16}$ inch steel rod, passing entirely around the tray; painted one coat of heavy paint.

Price, $4.00.

Bolted Wooden Canal Barrow with Steel Wheel.

One of the strongest canal barrows made. The front of tray is supported by a wooden cleat, bolted as well as by the heavy iron braces; handles and legs of hard wood. Tray full size with edges shaved; steel wheel that will not fall to pieces like the common wooden wheel.

Price, $2.00.

Henderson's "All-about" Hand Carts.

Strong and durable, adapted for a great variety of uses. They are handy about the orchard, garden, farm, lawn and stable for carrying tools, vegetables, fruits, etc. All finely painted, striped and varnished.

☞ The undermentioned prices are for Carts without springs; if springs are wanted, add $1.50 per cart extra. Any of above carts can be furnished with a third wheel in front, in place of the iron rest, for $1.50 extra.

No. 1, 36 in. wheels, Box 40x23 in., 10 in. deep, 1 in. tire, $8.00; 3 in. tire, $10.00
No. 2, 30 " " 32x20 " 9 " 1 " 7.00

The Lawn and Garden Hand-Cart.

A most convenient cart. Is far superior to any wheelbarrow for all kinds of work.

Having two wheels, it is self-supporting when in motion, and the operator does not have to hold it from turning over sidewise.

It has large wheels, 32 inches in diameter, which are placed well under the box, so that the wheels carry nearly all of the load instead of the man carrying about one-half of it. Box is 3 feet 10 inches long, 21 inches wide inside bottom, and 15 inches deep.

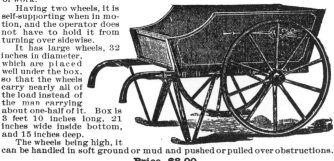

The wheels being high, it can be handled in soft ground or mud and pushed or pulled over obstructions.

Price, $8.00.

N. Y. PATTERN HORSE-RADISH GRATER
ON LEGS.

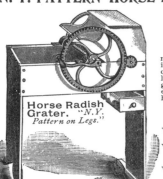

Horse Radish Grater. *"N.Y. Pattern on Legs."*

A very popular machine for grating horse-radish, cocoanut and similar substances; will grate finely and evenly 20 lbs. per hour.

Weight, 75 lbs.

PRICES.

With perforated tin cylinder, $8.00

With wooden cylinder steel pins, . . $9.00

With Treadle Attachment extra, . . $3.00

Watts' Asparagus Buncher.

THE BEST BUNCHER MADE.

May be adjusted to any size bunch by simply loosening the bolts at either end and pulling out the arms to fit; thus both ends of grass can be bunched properly. When the grass grows larger at one end than the other this will be found a great advantage, as either end can be regulated.

PRICE, $3.00.

BOX PATTERN HORSE-RADISH GRATER.

This machine is small, compact and very convenient to use. Can be operated by one person. The cylinder is covered with heavy perforated tin. Will grade about 15 lbs. of horse-radish or cocoanut an hour.

Weight, 32 lbs.

PRICES.

With perforated tin cylinder, $6.00

With wooden cylinder studded with steel pins, . . $7.00

Steel Carriage Jack.

No cast metal. No wood. Strong, light and compact.

Double lift bar operated by powerful compound levers.

Adjustment rapid and convenient, with wide range in height.

PRICES.

No. 1, to lift 900 lbs., $1.50 each.

No. 2, to lift 1,800 lbs., $2.00 each.

No. 3, to lift 4,000 lbs., $4.00 each.

Steel Bi-Treadle
FOR GRINDSTONES.

You sit down and work the treadles with both feet; your hands are free to attend to your sharpening. It runs almost as easy as a sewing machine; has a self-watering attachment. Not necessary to call a boy to turn the stone for you now.

PRICES.

Without stone, $4.00.
With   "   5.50.

The Farmer's and Stockman's
COMPLETE KIT OF
BLACKSMITH'S TOOLS.

Every tool the best made.

COMBINATION ANVIL and VISE, solid and strong. Weight, 55 lbs. Face 4¾ x 9 inches; jaws, 3 inches wide and open 4 inches. $6.00.

DRILL, a genuine blacksmith's post drill, with adjustable table. Drills ¾-inch hole to the centre of a 17-inch circle. $8.00.

FORGE, 15 inches high to top of bowl; bowl 14 inches in diameter; fan 8 inches in diameter; light running, strong blast; will heat 1½-inch iron. $10.00.

SCREW PLATE, 3 taps and 3 sets dies, cutting ½, ⅜ and ¾ inches.

BLACKSMITH'S TONGS. CAST-STEEL PINCERS, 12 inches long.

BLACKSMITH'S HAMMER, 2 pounds. ADZE-EYE SHOEING HAMMER, 9 ounces.

FARRIER'S KNIFE (Woostenholm's).

BLACKSMITH'S HOT CHISEL HAMMER, 1½ pounds.
    "   COLD   "   "   "

We warrant every tool the best made and the kit complete the best and cheapest on the market. Satisfaction guaranteed.

Price, complete, $25.00.

THE "PLUMLEY" FRUIT PICKER.

A very simple device, without springs or machinery to get out of order or injure the fruit, which is pulled off by three curved iron fingers, and it rolls easily down a cloth tube, the bottom of which can be held in a basket or barrel. By holding the cloth tube with one

THE "PLUMLEY" FRUIT PICKER.

hand the fall of each fruit is checked and can be let out as easily as the operator desires, preventing bruising. As the picker does not have to be lowered with each fruit, as with some pickers, the result is ten times as much can be picked, and with greater ease, by one person. The length of the tube is 11½ feet, enabling a man of ordinary height to reach fruit 16 feet from the ground, and as much lower as desired.

PRICE (without pole), **$1.00.**
    "   (with pole),   **1.50.**

SAMSON GRINDSTONES.

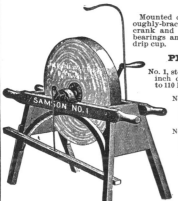

SAMSON NO. 1

Mounted on strong, thoroughly-braced frames, with crank and treadle, roller bearings and hook for the drip cup.

PRICES.

No. 1, stone about 24-inch diameter, 100 to 110 lbs., . . $4.00.

No. 2, stone about 22-inch diameter, 70 to 80 lbs., $3.50.

No. 3, stone about 18-inch diameter, 40 to 50 lbs., $3.00.

THE VICTOR PLATFORM SCALE.

An excellent scale, mounted on wheels; finished in natural wood, with brass beam and brass sliding poise.

PRICES.

No. A, platform 16 x 22 inches; capacity 500 lbs., $12.00.

No. B, platform 17 x 26 inches; capacity 1,000 lbs., $15.00.

"AUTOMATIC" ORCHARD Step-Ladder.

A NEW ERA IN STEP-LADDERS; IT STANDS WHERE OTHERS FALL.

It is virtually a four-legged tripod, and adjusts itself to uneven ground. It does not require careful adjustment before you step up, for it always maintains its equilibrium, even if one leg sinks unexpectedly, as it instantly adjusts itself to the new position. A light but very strong, safe and durable bolted ladder with malleable, double-acting hinge head.

PRICES.

6 feet, . . . $1.75.
8   "   . . . 2.25.
10   "   . . . 2.75.

DWELLING AND GREEN HOUSE FORCE AND LIFT PUMP.

A first-class pump, with all improvements, easy working, powerful and durable. Adapted to lifting water from cisterns or wells not exceeding 25 feet in depth. It is used both as an ordinary lift pump, discharging water from the spout, or hose may be attached to the latter and the water forced through it, for washing windows, watering gardens and numberless other purposes. The spout is threaded for 1-inch hose coupling ; if your hose is the regulation garden size (¾-inch bore), a 1 x ¾-inch reducer—cost, 25 cents—will connect hose and spout. Or, if preferred, the second discharge from top of air chamber may be utilized by unscrewing the cap closing it, in place of which permanent fittings of 1¼-inch pipe may be connected. These pipes or pipe may lead off along the benches of greenhouse or to tubs or upper floors of dwelling, wherever required. To force water through these pipes simply close the cock in the spout.

In addition to the pump you will require suction pipe to reach from the platform on which the pump is to be fastened to nearly the bottom of the well. Price of pipe extra :
1 inch, 6 cents per foot ; or galvanized, 8½ cents per foot ; 1¼ inch, 8 cents per foot ; or galvanized, 11 cents per foot.

Suction or strainer basket for bottom of pipe, if desired, 50 cents more. Fittings for pipe discharge from top of air chamber will be quoted on application, accompanied by diagram and measurements.

PRICES AND SIZES.

No.	CYLINDER.	CAPACITY PER STROKE.	SUCTION.	IRON.	BRASS CYLINDER.
0	2 inch.	.08 gallons.	1-inch pipe.	$7.50	$11.00
2	2½ "	.13 "	1¼ " "	8.50	12.00
4	3 "	.18 "	1¼ " "	10.00	13 00

THE AËRATING AND PURIFYING WATER ELEVATOR
FOR WELLS AND CISTERNS.

CLEAN, HEALTHY, COLD WATER DELIVERED IMMEDIATELY.

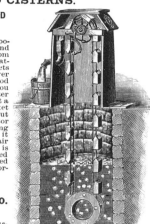

The buckets go into the water in an inverted position, full of air, which escapes gradually, and when the buckets turn upwards around the bottom reel the air is all discharged, thus thoroughly aërating and purifying the stagnant water. The buckets fill from the lowest and coolest depth and deliver it at the curb free from the taste of rotten wood tubing, rusty iron or "rain water" taint. You don't have to pump half a barrel of tepid water to get a cool drink in the summer, nor thaw out a frozen iron pump, nor kick the old oaken bucket from its icy lodgment in winter. No worn-out valves to pack, for the Aërating Water Elevator works just the same, winter or summer, elevating water rapidly and so easily a child can operate it and anybody can set it up, take it down or repair it, for it is as simple as a latch string. The curb is handsomely painted and decorated ; double-geared fixtures ; the buckets and chain are galvanized to prevent rusting. Everything about it is thoroughly first class and guaranteed.

PRICE, Complete, with sufficient chain
for a well 10 feet deep, . . **$10.00.**
Extra bucket chain, per foot, 25 cents.

To determine the length of chain required, measure the distance from the bottom of the well to a point two and one-half feet above the platform, and double it.

Garden Engine.

Very useful about lawns and gardens for watering plants, grass, washing windows, wagons, etc. It forces water from 30 to 50 feet from the nozzle, according to whether a sprinkling or stream tip is used. The tank holds about a barrel of water (40 gallons), and being on wheels, is convenient for moving about.

We furnish nozzle and 3 feet of discharge hose with the engine, but a 25 or 50-foot length of hose can be used, if desired. If you have the regulation garden hose (¾-inch bore), we can supply a 1 x ¾-inch reducer for 25 cents, so it can be connected with the engine hose.
PRICE, $18.00.
Or, if fitted for side suction, so the tank may be filled from a stream or pond, the price will be $20.00, or with 6 feet of 1½-inch suction hose and strainer basket, as shown in the cut, $25.00.

Portable Agricultural Pump.

A strong, well-made pump with brass cylinder and boxes, of large capacity. Will pump a barrel or more a minute. It is mounted on a wrought-iron tripod, the legs of which fold together, for carrying.

Every farmer should have one of these pumps to fill water carts or other receptacles from stream or pond. The improved valves especially adapt it for pumping up liquid manure.
PRICE, 4-inch bore, 8½-inch stroke, with 6 feet 2-inch suction hose and suction strainer, . . **$28.00.**

Farmers' DOUBLE-ACTING Tank Pump.

Has a capacity of from 1 to 1½ barrels per minute. It will lift water 25 feet from a stream or pond. It will fill a wagon tank or any other receptacle with water from a stream or pond, or manure water from the barnyard at the rate of a barrel or more a minute. After filling, place the suction hose in the tank and the water can be forced through the discharge nozzle for a distance of about 50 feet, rendering this pump of great value in distributing water on crops or grass lawns, etc., and is largely used as a thrasher tank pump.

AGRICULTURAL PUMP.
PRICE, with 15 ft. 2-in. spiral wire suction hose and suction basket, 12½ ft. 1-in. discharge hose and nozzle, couplings, etc., complete, $25.

"EMPIRE" DOUBLE-ACTING FORCE AND LIFT WELL PUMP.

The highest development in well pumps, thoroughly good in every respect. It is adapted for wells open or drilled, shallow or deep, not exceeding 100 feet. It throws a uniform, continuous and powerful stream and "pumps easier" than any other pump of same capacity. Being double-barreled, it raises a given quantity of water one-half easier than it can be raised in ordinary single-barreled pumps, for in the latter the whole lift comes on the down sweep of the handle, while in the "Empire" double barrel it is distributed over both the up and down sweeps, and you lift only one-half the quantity of water at each time. The spout is arranged so that ¾-inch hose can be screwed on, adapting it for a perfect force pump, for washing wagons, windows, watering gardens, etc. It is made for use and durability ; brass-lined cylinder and differential plunger. Glands and stuffing boxes are dispensed with, avoiding friction and wear.

PRICE, $10.50.
(Manufacturer's Price, $14.00.)
In addition to the pump you will require 1¼-inch suction pipe to reach to the bottom of your well. The pump, as offered above, reaches 4½ feet below the platform on which it is fastened ; therefore, if the bottom of the well is 25 feet below the platform, you will require 20 feet of 1¼-inch pipe, which we will supply at 8 cents per foot, or galvanized at 11 cents per foot.
Suction or strainer basket for bottom of pipe, if desired, will be 50 cents more.

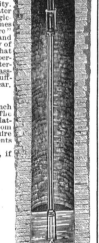

FARMERS' TANK PUMP.

Wheelbarrow Force Pump.

Another very useful machine about lawns and gardens where there is a pond or a stream of water or an accumulation of manure water. It will lift water 20 feet and force it through the discharge hose 30 to 50 feet from the end of the nozzle, according to whether a spray or stream tip is used on the nozzle.

We supply 3 feet of 1-inch discharge hose with it, to which you may attach a 25 or 50-foot length, if desired. Should you have the regulation ¾-inch bore garden hose, then we can supply you a reducer for 25 cents, so the two sizes may be connected. It will be found valuable for watering lawns and gardens, washing windows, wagons, walks, etc.
PRICE, complete, with 6 feet of 1¼-inch suction hose, strainer basket, 3 feet of 1-inch discharge hose and nozzle, . . **$16.00.**
For extra length of suction hose add 75 cents per foot.

GALVANIZED SHEET IRON HAND SUCTION PUMP.

The entire suction barrel being constructed of but one piece of galvanized iron spiral tubing, without joints, the superiority of these pumps is apparent.

Size.	Price.	Size.	Price.
Inside diam.	*Per foot.*	*Inside diam.*	*Per foot.*
1½ inches	$0.55	3½ inches	$0.75
2 "	.60	4 "	.80
2½ "	.65	4½ "	.90
3 "	.70	5 "	1.20

Less than 5 feet charged as 5 feet.

Magic Ham-Curing Pump

Cures hams during any season of the year quicker and better than by the old process. It impregnates them with a pickle preparation, recipes for which will be sent to all purchasers. All working parts of the pump are brass. We furnish it complete with 3 feet of ½-inch suction and discharge hose, brass suction strainer and nickel-plated needle for **$10.00.**

HAND SUCTION PUMP.

GALVANIZED STEEL WIRE GARDEN TRELLIS.

For PEAS, TOMATOES, VINES, Etc.

A great improvement in Garden Trellis, indispensable in every well-kept garden, easily put up, more tidy than brush, practically indestructible. Can be rolled up, stored away and used again year after year. Tomatoes grown on this Trellis are clean, ripen more evenly and are less liable to rot. Cucumbers and other vines can be advantageously trained on the Trellis, economizing space and insuring cleaner and better matured fruits. This Trellis, 46 inches wide, is made of the best galvanized wire, and is furnished in 10-foot lengths, with a stake at each end and one in the centre. **Price**, per 10-foot length, **75c.**; per doz. lengths, **$8.50.**

THE RANDALL WOVEN GALVANIZED STEEL WIRE FENCING.

The best wire fencing made; the only kind we use on our farms.

For PRIVATE GROUNDS, PARKS, ORCHARDS, GARDENS, CEMETERIES, and FARMS.

The meshes are connected by a series of loops without tendency to enlarge one way and contract another way under strain, while the top and bottom are made of 3-ply cable, smooth, uniformly laid and exceptionally strong. This Fencing always maintains its shape, and keeps taut under all conditions of heat and cold; it does not sag or buckle between the posts.

PRICES FOR NETTING ALONE:

No. 1—36 in. high, per roll of 12 rods (*about 200 feet*), 65c. per rod, $7.80. (Weight, 130 lbs.)
No. 2—43 in. "   "   "   "   "   "   75c. "   9.00. ("   140   ")
No. 3—51 in. "   "   "   "   "   "   85c. "   10.20. ("   150   ")

SUPERIOR GALVANIZED RIDDLES.

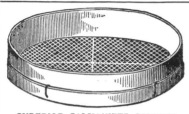

Do not rust; the wires cannot slip so as to form irregular openings, hence a perfect separation is always obtained; the wire cannot be pulled from the rim, and will not sag in the middle. The numbers indicate the number of openings to an inch. No. 2 for gravel, etc.; 3 beans, coarse sand, etc.; 4 corn and mason's sand, etc.; 5 and 6 fine sand, etc.; 8 wheat, rye, barley, etc.; 10 and over for special purposes. They are all 18 inches in diameter. **Price, 50c.** each, **$5.00** per doz.
Special Oat Sieves (No. 12), for taking dust out of oats before feeding. **50c.** each, **$5.00** doz.
Extra Heavy Florists' Riddle, for rubbing moss, woods, mold, cocoanut fibre, and lumpy loam through. ½, ⅝, ¾, ⅞ and 1-inch mesh; 18 inches diameter. **$1.50** each.

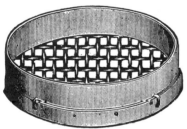

EXTRA HEAVY FLORISTS' RIDDLE.

HENDERSON'S SUPERIOR GALVANIZED WIRE NETTING,

For Poultry Yards, Fencing, Garden Borders, Etc.

Our netting is made of high-tempered steel wire of maximum strength; it is heavily galvanized with zinc after it is woven, which renders it "good for 100 years," and unites the wires together; the meshes are regular; it is woven straight and does not bulge. Warranted full measure. No. 19 wire, 2-inch mesh, is the popular poultry netting, though we can supply a heavier grade of wire at an extra cost, if specially wanted.

We do not sell LESS THAN ROLLS 150 FEET LONG.

NET PRICE PER ROLL 150 FEET LONG.

We can supply Netting in the under-mentioned widths:	3-inch Mesh. (No. 18 Wire.)	2-inch Mesh. (No. 19 Wire.)	1½-inch Mesh. (No. 19 Wire.)	1¼-inch Mesh. (No. 19 Wire.)	1-inch Mesh. (No. 20 Wire.)	¾-inch Mesh. (No. 20 Wire.)
12 inches wide..........	....	$1 50	$2 10	$3 00	$3 30	$5 40
18 " "		2 25	3 15	4 50	4 95	8 10
24 " "	$2 70	3 00	4 20	6 00	6 60	10 80
30 " "	3 38	3 75	5 25	7 50	8 25	13 50
36 " "	4 05	4 50	6 30	9 00	9 90	16 20
42 " "	4 73	5 25	7 35	10 50	11 55	18 90
48 " "	5 40	6 00	8 40	12 00	13 20	21 60
60 " "	6 75	7 50	10 50	15 00		
72 " "	8 10	9 00	12 60	18 00		

☞ **WE DO NOT SELL LESS THAN ROLLS 150 FEET LONG.** ☜

HANDY SCREENS,

FOR ASHES, COAL, SAND, GRAVEL, Etc.

Extra heavy wire. Spruce frames. State whether you want ¼, ⅜, ½, ¾, or 1-inch mesh.

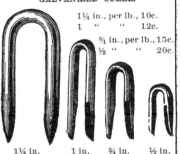

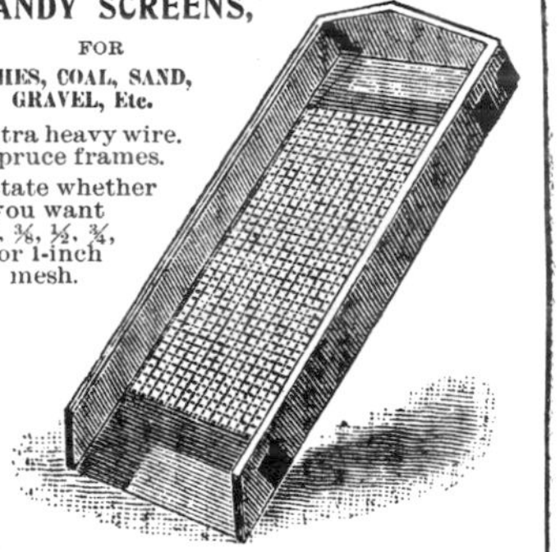

Small size, 25x62 inches, **$5.00**
Large " 28x66 " **6.00**

NETTING AND FENCE STAPLES.

GALVANIZED STEEL.

1¼ in., per lb., 10c.
1 " " 12c.
¾ in., per lb., 15c.
½ " " 20c.

1¼ in.   1 in.   ¾ in.   ½ in.

IRON SWEET PEA AND FLOWER TRELLIS.

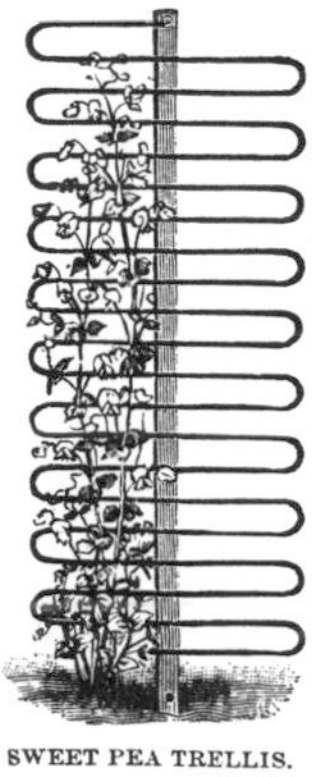

A most convenient, strong, and durable iron trellis, suitable for sweet peas or other garden flowers. An iron wing on the portion going into the ground keeps the trellis as steady and upright as placed. This trellis measures 24 inches wide, 52 inches long, 12 inches of which goes in the ground. **Price, 35c.** each, **$3.50** doz., **$28.00** per 100.

SWEET PEA TRELLIS.   TOMATO SUPPORT.

TRIPOD TOMATO SUPPORTS.

GALVANIZED IRON.

The tripod-shaped support is very strong, and will not blow over nor crush down. Tomatoes supported by these affairs ripen earlier, more evenly, and better than when allowed to lie on the ground, or even than when tied to stakes, for with this support the vines are open, admitting air and sunlight. **Price, 20c.** each, **$2.00** per doz., **$15.00** per 100.

"BEST STEEL"

COMBINED WIRE PLIERS AND CUTTERS.

4½ in., 60c.; 6 in., 75c.; 8 in., $1.00; 10 in., $1.50.

TACKLE-BLOCK WIRE STRETCHER.

This tool is a complete stretcher for all kinds of wire fencing, whether netting or barbed wire, and is a convenient set of pulley blocks besides, the whole costing no more than is ordinarily paid for one separately. All the iron is either malleable or wrought, and 15 feet of manila rope is furnished with each set. The pulleys are 2 inches in diameter. They are strong enough for 1,000 lbs. strain. The rope grapple is very convenient and quickly adjusted. They give better satisfaction than any wire-stretchers ever introduced. **PRICE $1.25**

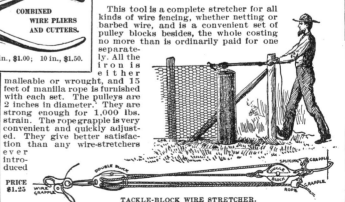

TACKLE-BLOCK WIRE STRETCHER.

Henderson's Hot Bed and Cold Frame Mats.

To Protect Plants from Frost in Winter and Spring.

"BURLAP" MATS.

These are made of strong burlap bagging, warmly lined with waste wool and cotton, which is quilted in to hold position. They are excellent substitutes for straw mats, being, if anything, warmer than straw, more easily handled, less bulky, and they do not harbor mice or other vermin. We were afraid that they would hold moisture, and either rot or mildew, but our trial for two winters proves them to be far more durable than straw mats.

PRICES { Size A, 40 x 76 inches, 60c. each; $6.00 per doz.
" B, 76 x 76 " 90c. " 9.00 "

"WATERPROOFED DUCK" MATS.

These are filled with cotton and wool waste, and quilted the same as the "Burlap" mats described above. The "*waterproofed duck*" cotton cloth on one side sheds water, and prevents them from getting "soaked through."

PRICES { Size C, 40 x 76 inches, $1.25 each; $12.00 per doz.
" D, 76 x 76 " 1.75 " 18.00 "

RYE STRAW MATS.

Being made in the best manner of long rye straw and best tarred cord, they are invaluable for throwing over cold frames, hot beds, etc., during the coldest weather; they roll up and can be stowed in small space.

PRICES { Size, 3 x 6 feet, $1.25 each; $14.00 per doz.
" 6 x 6 " 2.00 " 22.00 "

TEMPORARY CHRYSANTHEMUM HOUSE OF PROTECTING CLOTH.

PATENT PROTECTING ··· CLOTH. ···

Specially prepared to prevent mildewing and rotting; valuable for protecting plants from frost, covering hot beds and frames in spring, in lieu of glass, for Chrysanthemum houses, for covering tender bedding plants at night when there is danger of frost, thereby lengthening the display, etc., at one-tenth the cost of glass. It comes in yard widths.

PRICES.

Heavy Grade Protecting Cloth. Per yd., 12c.; per piece of about 40 yds., at 11c. per yd.; weighs about 46 lbs. per 100 yds.

Medium Grade Protecting Cloth. *Best for general purposes.* Per yd., 10c.; per piece of about 60 yds., at 8½c. per yd.; weighs about 25 lbs. per 100 yds.

Light Grade Protecting Cloth. *Mostly used in South for tobacco plants.* Per yd., 5c.; per piece of about 68 yds., at 4c. per yd.; weighs about 7¼ lbs. per 100 yds.

COLD FRAME WITH PROTECTING CLOTH IN LIEU OF GLASS.

MARKET GARDENERS' AND FLORISTS' SASH.

We carry a large stock, so that we can fill almost any order on receipt. **Unglazed.**—3 x 6 feet, for 6 x 8-inch glass, of best cypress, 80c. each; $9.00 per doz.; $70.00 per 100. This is the regulation sash for hot beds and cold frames, and we sell thousands of them every year.

EXTRA FINE SASH FOR PRIVATE PLACES.

Made from "red gulf" cypress, finished edges, with neat iron crosspiece running through the centre; painted one coat. **Unglazed.**—3 x 6 feet, requiring three rows of 8 x 10-inch glass, $1.25 each; $14.00 per doz. **Glazed.**—3 x 6 feet, with three rows of 8 x 10-inch glass, painted two coats white, $2.75 each; $32.00 per doz.
☞*For shipping, glazed sash has to be carefully crated. This will cost extra, 50c. per crate. Up to 6, glazed sash can be put in a crate.*☜

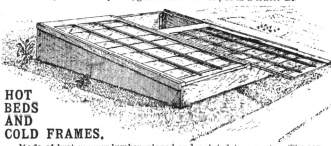

HOT BEDS AND COLD FRAMES.

Made of best cypress lumber, planed and painted two coats. The corners join by bolting on angle irons. The rails between the sash are rabbetted for the sash, grooved for drip and dove-tailed into the sides. Strong, durable and portable. Can be taken down for convenient storage when not in use; 8 inches deep in front, 16 inches deep at back.

Prices DO NOT include sash (for prices of sash, see above).

No. A—For two 3 x 6-ft. sash, $7.50; **No. C**—For four 3 x 6-ft. sash, $10.50
No. B— " three " 9.00; **No. D**— " five " 12.00

DEEP COLD PITS,

For wintering bulbs, carnations, dormant roses, etc.
Above ground they are the same as the hot-bed frames offered above, and, in addition, have posts and 2-inch lumber for siding up the excavation, 18 inches below the surface.

Prices DO NOT include sash (for prices of sash, see above).

No. E—For two 3 x 6-ft. sash...$10.50
No. F— " three " " 12.50
No. G— " four " " 14.50

Bryant's Plant Protector.

Protects young plants from insects, frosts, cold winds, etc., and thereby forwarding their growth in early spring. The bows are of bent wood covered with mosquito netting.
PRICES.—15c. each; $1.60 per doz.; $12.00 per 100.

Waterproof Paper Plant Protector.

Will last for years. Cone-shaped, for individual plants.
6 ins. high, $0.25 pr.dz.; $1.75 pr.100
9 " " .50 " 3.75 "
12 " " .75 " 5.50 "
15 " " 1.00 " 7.75 "

"Pot-Lid" or Hill Protectors.

For vines and seedlings in masses.
10 in. diam.. $0.50 pr.dz.; $3.75 pr.100.
14 in. " .75 " 5.50 "

WATERPROOF PAPER PLANT PROTECTORS.

MASTICA. For glazing Greenhouses, Sashes, etc., new and old. It is elastic, adhesive, and easily applied. It is not affected by dampness, heat or cold.

MASTICA GLAZING MACHINE.

Every florist has experienced difficulty in obtaining putty (whether ordinary or white lead) for glazing that is satisfactory for any length of time. The fact is, putty is not adapted for greenhouse work. "Mastica" is elastic and tenacious, admitting of expansion and contraction without cracking. When applied a few hours, it forms a skin or film, hermetically sealing the substance and preventing evaporation of the liquids, and remains in a soft, pliable and elastic condition for years. "Mastica" is of great value in going over old houses with a mastica machine on the outside of sash, as it makes it perfectly tight and saves the expense of relaying the glass. One gallon is sufficient for about 300 lineal feet, used either for bedding or over glazing. PRICES for Mastica, soft, for machine application. 50c. per quart; 75c. per ½ gallon; $1.25 per gallon. **Mastica Glazing Machine, $1.25 each.**

FLORISTS' DIAMOND GLASS CUTTER.

Better than any other cutter, and works entirely different, having a diamond in one corner and a wheel in the other, thus giving a sure cut. Will never lose the point.
PRICES.—Cocoa handle (*the diamond is nearly twice the size of the one in the ebony handle*), $4.00; ebony handle, $3.00.

Van Reyper's "Perfect" Glazing Points.

The finest glazing points on the market. These are made of steel wire and galvanized, having double points, and lap over the glass in such a manner as to positively keep it from slipping. They may be used on either side of the sash bar, thus preventing the annoyance of rights and lefts.
PRICE, per box of 1,000, 65c. Pincers, per pair, 50c.

VAN REYPER'S GLAZING POINT.

FLORISTS' DIAMOND GLASS CUTTER.

ASPARAGUS KNIVES.

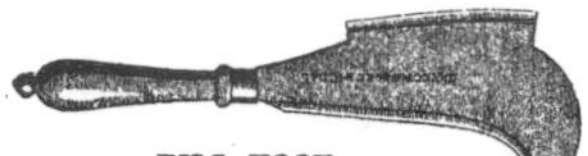

BILL HOOK.

BULL LEADER.

BULL PUNCH.

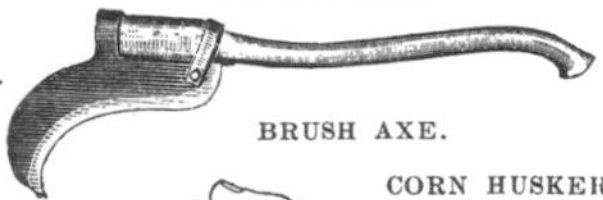

BRUSH AXE.

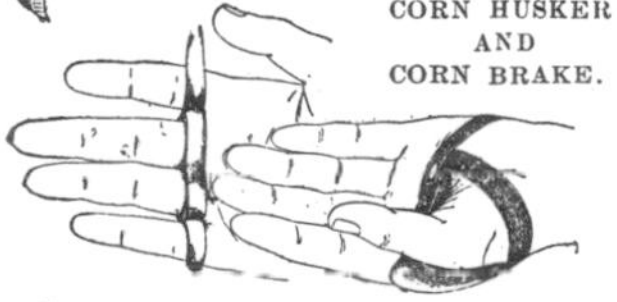

CORN HUSKER AND CORN BRAKE.

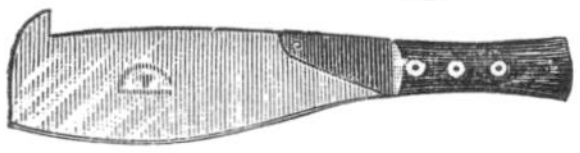

MACHETE CANE AND CORN KNIFE.

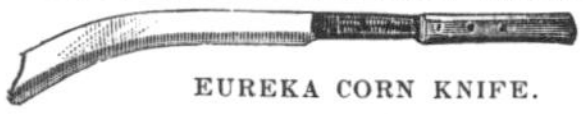

EUREKA CORN KNIFE.

SERRATED EDGE CORN KNIFE.

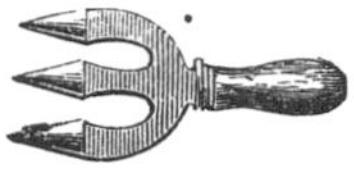

HAND WEEDING FORK.

DIGGING FORK.

MANURE FORK.

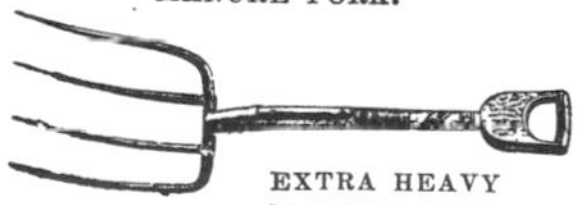

EXTRA HEAVY MANURE FORK.

HAY FORK.

VEGETABLE SCOOP FORK.

WOODEN STABLE FORK.

STONE PICKING FORK.

FLORAL TOOLS.

GRAFTING CHISEL.

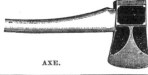

AXE.

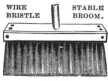

BURNING BRAND.

SEAMLESS STEEL PATENTED BUSHEL BASKET

WIRE BRISTLE   STABLE BROOM.

HORSE BRUSH.

CURRY COMB, WITH MANE COMB.

CHAIN COW TIE.

DIBBER.

DRAIN CLEANER.

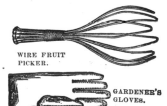

THE PERFECT FRUIT PICKER.

WIRE FRUIT PICKER.

GARDENER'S GLOVES.

GARDEN LINE AND REEL.

FISCHER'S HAY KNIFE.

LIGHTNING HAY KNIFE.

	Price.
Asparagus Knife. Straight edge, 25c. each..................................doz.,	$2.50
slant edge, 30c. each, $3.00 doz.; V-shaped edge, 35c. each........doz.	3.50
Axes. Best quality steel, all handled—**Light,** 2½ to 2¾ lbs., 80c.; **Medium,** 3 to 3¼ lbs., 90c.; **Heavy,** 3½ to 4 lbs., $1.00; **Boy's Axe,** handled...	75
Axle Grease, etc. (*See under Lubricants.*)	
Baskets—Seamless Steel Bushel Baskets, pressed from one piece of steel; no joints or seams; do not leak or rust; being galvanized, may be used for solids or liquids...each,	2.40
Farmer's Corn Baskets, Best Woven Oak Splint. Peck, 30c. each; ½ bushel, 45c. each; 1 bushel, 60c. each; 1½ bushels...each,	75
Extra Strong Woven Rattan Farm Baskets. 1 bushel, $1.00 each; 1½ bushels, $1.35 each; 2 bushels...............................each,	1.75
Market Gardener's Vegetable or Potato Baskets. Woven split reed, hole handles under rim. ¼ bushel, 45c. each; ½ bushel, 55c. each; 1 bushel...each,	65
Bill Hooks. For cutting out underbrush, trimming hedges, etc....each,	1.25
Bull Leader or Staff, $1.50; **Copper Bull Rings,** 2½ or 3 inch, each,	25
Bull Punch. For making hole in nose to insert ring........................	60
Burning Brand. Made to order with any name or initials for marking implements and tools, ¼ inch, ⅜ inch or ½ inch letters. Price of brand, with six letters or less, $1.50; additional letters..............each,	20
Brush Axe. Axe handled for shrubbery and underbrush......................	1.00
Brushes, Horse Brush. All bristle, leather backs, $1.00; ext. quality,	1.25
"Utility" Steel Brushes. 2 inch bristles of flexible steel wire; size of brush, 9x2¼ inches; fine for cleaning horses, cows, etc............each,	75
Brooms—Stable Brooms, Steel-wire bristles. Superior to rattan bristles; these keep their shape and last longer. 12 inch broom, 75c. each; 14 inch, 85c.; 16 inch...	1.00
Cattle Cards. Best quality, large size.........................per pair,	30
Curry Combs. Extra quality, 8 row, close back, with mane comb........	30
" " **"Circular."** Flexible spring; a perfect spring comb..	35
Chain Cow Ties. *"New Idea."* Smooth, strong and safe; made from solid steel wire, with steel rings, toggles and swivels; no welds...each,	25
Bull Size, extra strong...each,	40
Halter Chains. *"New Idea,"* with safety snap, swivel and slide ring; tested to 1,000 lbs.—4½ feet, 20c. each; 6 feet, 25c. each tested to 2,000 lbs.—4½ feet, 30c. each; 6 feet.....................each,	40
Picket or "Staking Out" Chains. Steel, smooth American link, with safety snap and swivel on one end and steel ring and swivel on the other end. 20 feet, $1.00; 30 feet......................................	1.50
Chain and Tether Stake. Wrought-iron stake, with swivel on top; with twenty feet of chain, with swivel in chain...........................	1.50
Corn Husker. Excelsior Corn Husker; malleable iron; unbreakable; fits either hand. 15c. each.....................................doz.,	1.50
Corn Break, The Eureka. For breaking off ears in husking; fits either hand; twice as much can be done with it, 25c. each....................doz.,	2.50
Corn Knives, Serrated Edge. Finest steel; balanced handle is not point heavy, 40c. each...doz.,	4.00
Eureka Corn Knife. Forged from solid steel, 30c. each.........doz.,	3.00
"Machete" Corn and Cane Knife. Finest steel.....................	65
Cradles, Grain Cradles. The turkey wing pattern, used in northern states; 5 fingers; complete with scythe..................................	3.00
Crowbars. Steel, wedge pointed; 10 to 24 lbs.....................per lb.,	6
Dibbers. Henderson's Round Dibble for transplanting....................	35
Flat Steel Dibble for plants, 2½ x 9 inches........................	60
" " for nursery stock, 4 x 10 inches....................	75
Drain Cleaner. To push and pull; concave blade, 4 inches wide, 75c.; 5 inches wide, 80c.; 6 inches wide...................................	90
Drain Tile Layer..	75
Floral Tools. Small, for children; set of four pieces................	50
Large, long-handled, good material, useful for ladies..............	1.00
Floral and Gardening Tools. Youth's Set. Five pieces: spade, hoe, rake, trowel and hand-weeder; first-class in every particular..............	1.25
Forks—	
Digging or Spading. Best quality; strapped D handle. Four prong, 75c.; five prong, 90c. *Long handled at same prices.*	
Manure Forks. Best quality; strapped D handles. Four tine, 75c.; five tine, 85c.; 6 tine, $1.00. *Long handled same prices.*	
"Extra Heavy" Manure Forks. Diamond-shaped tines. Four tine, 90c.; five tine...	1.00
Hay Forks. Best quality; long strapped handles. Two tine.......... three tine, 45c.; four tine...	35 / 60
Potato Digging Forks. Expressly made for the purpose; six heavy round tines, points not too sharp; strapped D handle........	1.25
Vegetable Scoop Forks. Will load to the head without raising the points; strapped D handle; eight tine, $1.50; ten tine...........	1.75
Wooden Stable Forks. For handling bedding without danger of sticking animals; best hickory; 3 tine.................................	75
Stone Picking Forks. The handiest tool for the purpose...........	85
Hand Weeding or Transplanting Forks. American malleable, 15c. English steel..	15 / 50
Fruit Picker, The Perfect. Galvanized steel wire, attaches to pole of any length. Price, without pole.......................................	60
Wire, Ordinary Shape. Without pole..............................	25
Garden Lines. Best braided linen, 100 feet.........................	50
Garden Line Reels. Malleable...................................	50
Gardener's Gloves. Heavy Goat, for thorny plants............pair,	1.00
Rubber, with gauntlets, men's sizes, $1.50; ladies' sizes...........	1.35
Grass Hooks. (*See page 53.*)	
Grafting Chisel. Handled..	65
Grafting Wax. Trowbridge's, ¼ lb., 10c.; ½ lb., 20c.; 1 lb.........	30
Hay Knives, Lightning. For cutting down hay in the stack or bale; also for ensilage, dry fodder, etc..	1.00
Fischer's Patent..	75

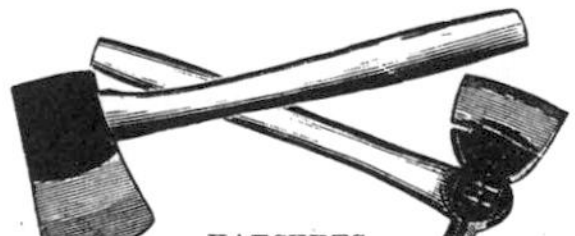
HATCHETS.

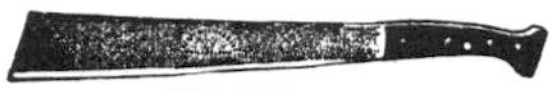
HEDGE KNIFE.

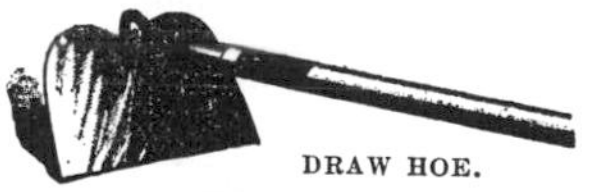
DRAW HOE.

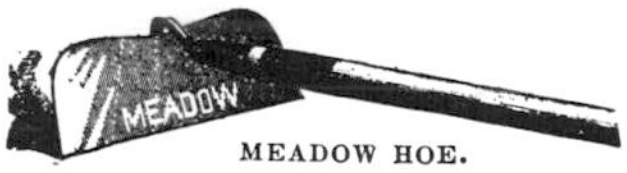

MEADOW HOE.

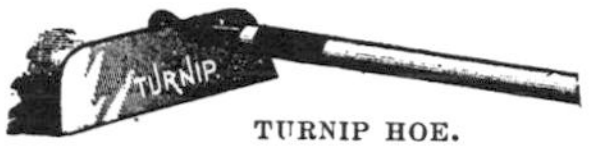

TURNIP HOE.

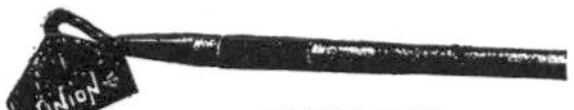

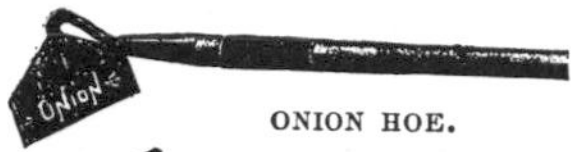

ONION HOE.

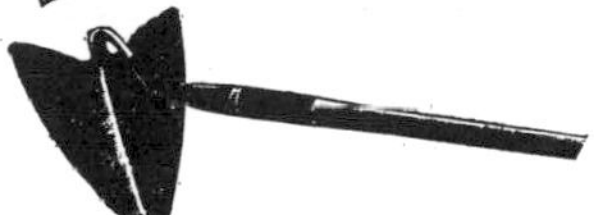
WARREN HOE.

ACME HOES.
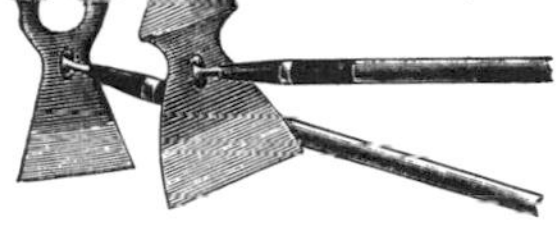

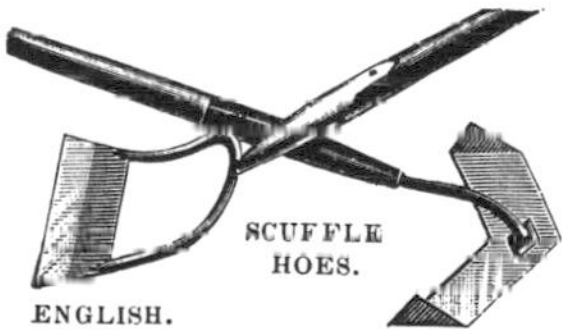
SCUFFLE HOES.
ENGLISH. LIGHTNING.

HALF-MOON SCUFFLE HOES.
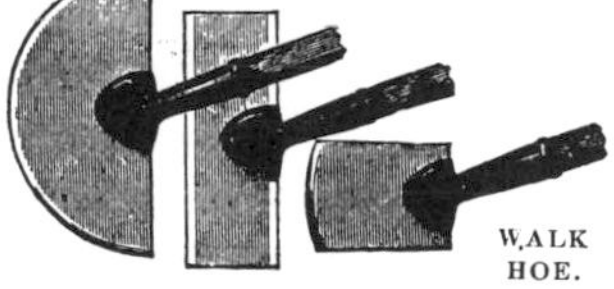
WALK HOE.

COMBINED SCUFFLE HOE AND RAKE.
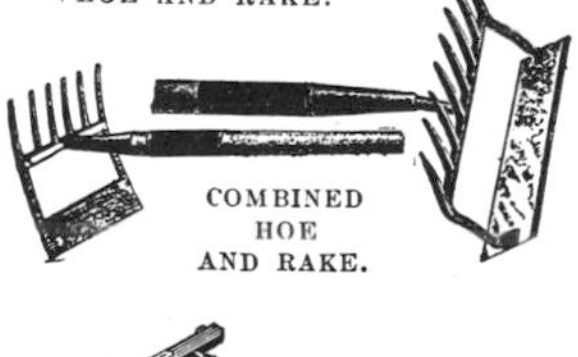
COMBINED HOE AND RAKE.

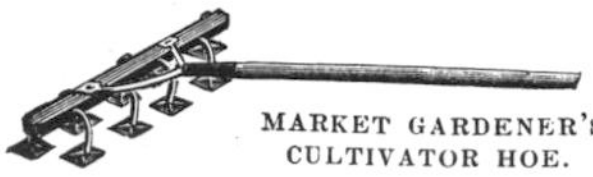
MARKET GARDENER'S CULTIVATOR HOE.

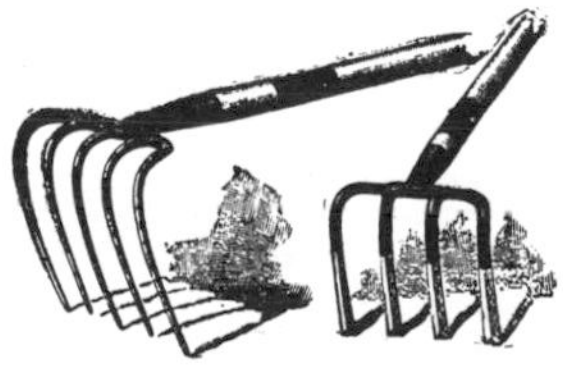
PRONG HOES, OR POTATO AND MANURE HOOKS.

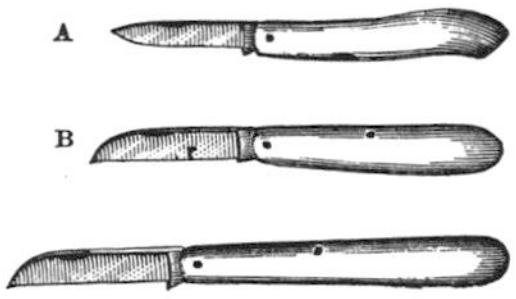

BUDDING KNIVES.

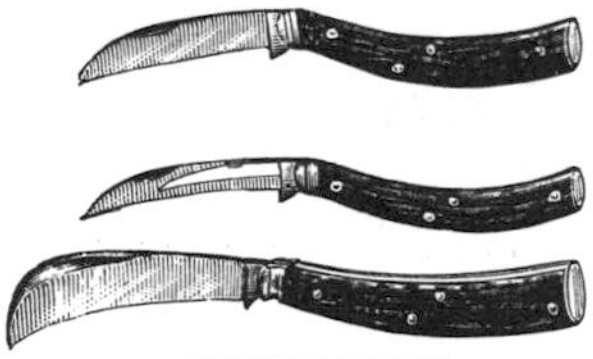

PRUNING KNIVES.

Price.

Hammers. Standard quality, 35c.; extra quality steel.......... 50
Hatchets—Axe Pattern. For pruning, hunter's use, etc. No. 1, 1½ lbs., 65
No. 2, 1¾ lbs., 75c.; No. 3, 2 lbs.......... 85
Shingling or Common Hatchet. 3½ inch cut, 50c.; 4 inch cut... 60
Claw Hatchet. 3½ inch cut, 65c.; 4 inch cut.......... 75
Barrel or Half Hatchet. 2¼ inch cut.......... 60
Broad or Bench Hatchet. 4 inch, 85c.; 5 inch, $1.10; 6 inch....... 1.35
Hedge Knife. 21 inches long, blade 13 inches; of finest steel for trimming woody hedges.......... 50
Heaters, Oil. For conservatories, bath rooms, etc.—
No. 10, 23 ins. high, holds two quarts oil, burns 9 hours.......... 3.50
No. 44, 32 " " " four " " " 18 " 6.00
Hog Rings. To prevent rooting, 10c. per doz..........per 100, 50
Hog Ringer. For attaching rings to nose..........each, 25
Hog Scraper. For removing hair after killing..........each, 30
Hoes—Draw or Field Hoe. Finest quality, socket handled, 5 inch, 25c. (*Ladies' Hoe*), and of the following sizes: 6 inch, 7 inch, 7½ and 8 inch, at 35c. each..........doz., 3.50
Meadow Hoe. Socket handled, 8 inch, 40c.; 9 inch, 45c.; 10 inch, 50
Turnip Hoe, 7½ inch, 40c.; 8½ inch.......... 45
Onion Hoe, 6 inch, 35c.; 7 inch, 40c.; 8 inch.......... 45
Warren Heart-Shaped Hoes. The finest garden hoe made. Useful for all purposes, hoeing, furrowing and covering; light and rapid to work with. Small size, 45c.; medium or general-purpose size, 50c.; large size.......... 55
Acme Weeding and Cultivating Hoe. Double Prong, 4½ inch blades, 50c.; **Single Prong,** 6 inch blades.......... 40
English Scuffle or Push Hoes. Imported. (Handles, extra, 10c. each.) 5 inch, 50c.; 6 inch, 55c.; 7 inch, 60c.; 8 inch, 65c.; 9 inch, 70c.; 10 inch, 75c.; 12 inch.......... 85
Lightning or V-Shaped Scuffle Hoe. Handled. Ends of blade turned up to prevent cutting plants. 8 inch cut.......... 70
Combined Scuffle Hoe and Rake. Handled. 8 inch cut and six-tooth rake, 75c.; 9½ inch cut and eight teeth.......... 85
Combined Draw Hoe and Rake. Handled; steel. 4¼ inch cut and four teeth, 35c.; 6½ inch cut and six teeth.......... 45
Grub or Cranberry Hoe.......... 1.00
Half-Moon Scuffle Hoe, 7½ inches diam., 50c.; 9 inches diam...... 60
Chisel Blade Walk Hoe, 3 inch blade, 35c.; 4 inch.......... 40
Market Gardener's Cultivator Hoe. A remarkable new tool; it works like a push hoe; cuts weeds and loosens the soil when moved either backward or forward; those who have used it are enthusiastic about the ease and quantity of work done with it and its thorough pulverization of the soil.
No. 1. Cuts 10½ inches........$1.25 No. 3. Cuts 6¼ inches 1.00
No. 2. " 8½ " .... 1.15 No. 4. " 4 " 85
Prong Hoes or Potato Hooks. 5 round tine, 40c.; 4 broad tine.. 40
Knives, Budding. Ivory handles, finest quality steel; straight blade, long handle (C), $1.00; straight blade, short handle (B), $1.00; round point blade, curved handle (A).......... 1.00
Knives, Pruning. Stag handles, finest quality. Medium size, single blade (D), 75c.; medium size, double blade (E), $1.00; large size, single blade (F).......... 85
Labels, Wooden. *Garden labels are put up in packages of 100; pot, plant and tree labels in packages of 500.*
Labels—

Garden,	8	inches, per 100..........	Plain, $0.40..........	Painted	50
"	12	" "	" 50..........	"	60
Pot,	4	" per 1,000..........	" 50..........	"	75
"	5	" "	" 75..........	"	1.00
"	6	" "	" 90..........	"	1.25
Wired,	3½	" "	" 1.50..........	"	1.75
"	2½	" "	" 1.25..........	"	1.50

Glass Label Holder. The best device for permanently labelling trees, shrubs, hardy roses, etc.; absolutely air and water tight; the glass is annealed and not liable to be broken. Write the name on card or paper, insert through the opening at bottom of holder, and seal with putty or plaster paris. Price, 5c. each, 50c. doz., mailed; or $3.50 per 100, buyer paying expressage.
Label Pencil, Indelible. Black lead, 5c. each..........doz. 50
Lanterns (*"Safety Tubular"*). The only lanterns burning kerosene that are safe to use in barns and around combustible material; they have the Stetson patent safety attachment; lighted, regulated, filled or extinguished without removing globe; will not blow out.
Square Frame Tin Safety Lantern, 65c.; copper-plated....each, 1.00
Round Frame, Tin, $1.00; brass house lantern..........each, 1.25
Driving Lamp (*The "Safety Tubular"*), with side or dash attachment. Japanned, $2.85; nickel-plated.......... 3.50
Large Square Lawn and Piazza Lamp (*The "Safety Tubular"*). Endures hardest storms without smoking or blowing out; no chimneys to break; will burn twelve to twenty hours.
No. 1. 1½ inch flame, 7 inch silvered reflector..........each, 5.00
No. 2. 3 " " 8 " " " ....each, 6.00
Hunting Lamp (*"Safety Tubular"*), with cap arranged to shut off light; tin, japanned black..........each, 5.00
Globe Hanging Piazza Lamp (*"Safety Tubular"*). Especially made for hanging in verandas, halls, etc.; no shadows; not affected by wind. Gives large, brilliant and steady light.......... 5.00
Mangers "Slowfeed." Cast-iron corner manger, 17 x 17 inches....... 1.60
Cast-iron corner hay-rack.......... 1.60
Measures, Seamless Steel. (*Dry Measures.*) Stamped from one piece, with heavy top band; galvanized. Quart size, 35c.; two quarts, 45c. four quarts, 60c.; eight quarts.......... 1.00
Oak Measures, ½ bushel, 75c.; peck, 45c.; four quarts, 35c.; two quarts, 25c.; one quart, 20c. **Nest** *of the above five sizes*.......... 1.65

HOG RINGER.
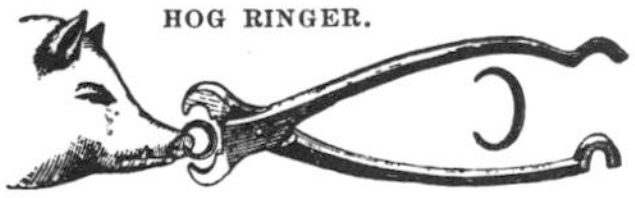

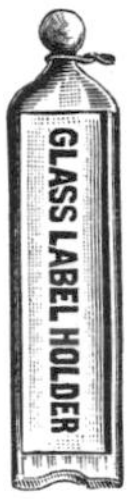

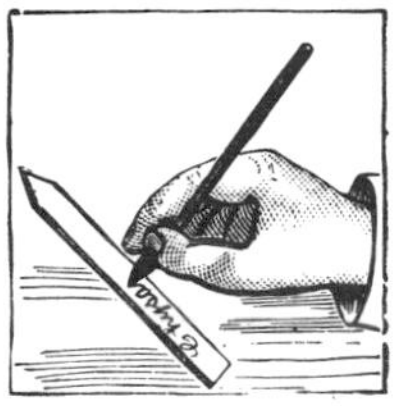
LABEL PENCIL.

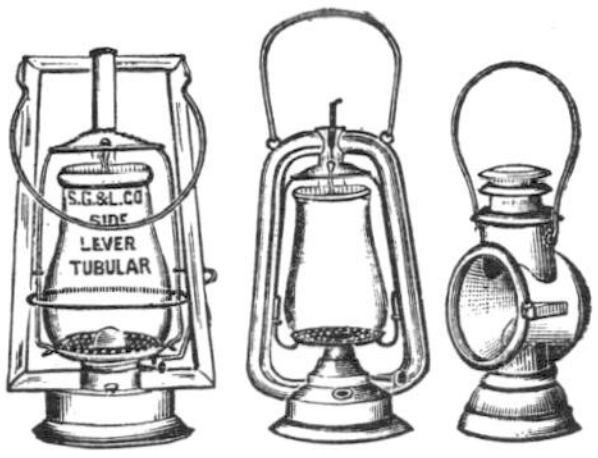

SQUARE FRAME. ROUND FRAME. DRIVING.
TUBULAR LANTERNS.

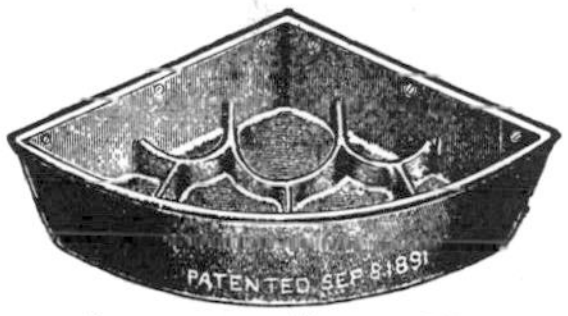

"SLOWFEED" MANGER.

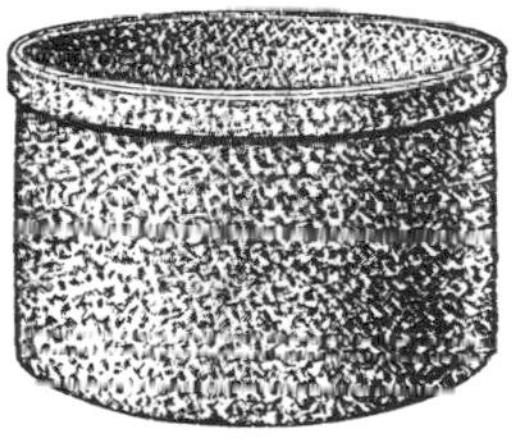
SEAMLESS STEEL MEASURE.

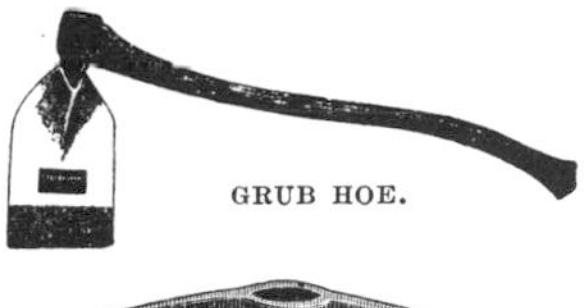
GRUB HOE.

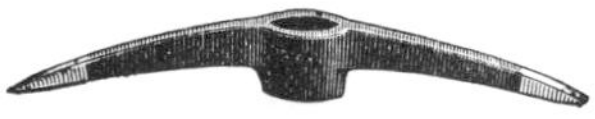
CLAY PICK.

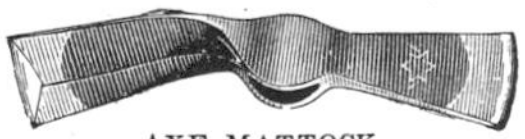
AXE MATTOCK.

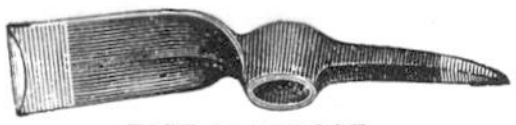
PICK MATTOCK.

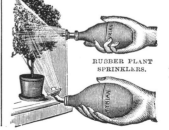

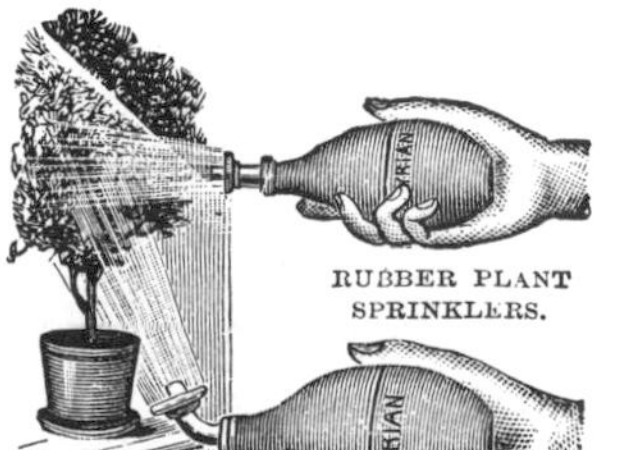
RUBBER PLANT SPRINKLERS.

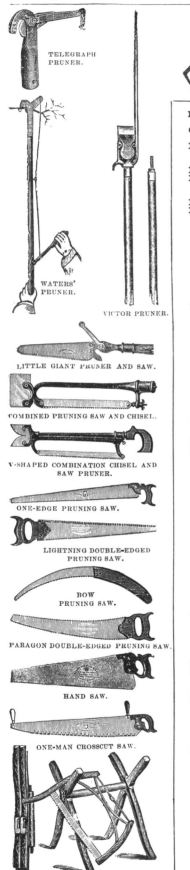

TELEGRAPH PRUNER.

WATERS' PRUNER.

VICTOR PRUNER.

LITTLE GIANT PRUNER AND SAW.

COMBINED PRUNING SAW AND CHISEL.

V-SHAPED COMBINATION CHISEL AND SAW PRUNER.

ONE-EDGE PRUNING SAW.

LIGHTNING DOUBLE-EDGED PRUNING SAW.

BOW PRUNING SAW.

PARAGON DOUBLE-EDGED PRUNING SAW.

HAND SAW.

ONE-MAN CROSSCUT SAW.

WOOD SAW AND SAW BUCK.

STEEL GARDEN RAKE.

STEEL BOW RAKE.

OLD-FASHIONED WOOD-HEAD GARDEN RAKE.

	Price.
Muzzles for Horses and Oxen. Used when cultivating corn, etc.; heavy galvanized wire......each,	25
Oil Cans. Spring bottom, squirt nozzle, 15c.; large mowing machine oilers......	40
Pails—Best Oak Stable Pail. Three galvanized hoops; large......	1.00
Indurated Fibre Ware Stable Pail. Light, everlasting, no shrinking, no hoops. 16 quart, 65c.; 18 quart, 75c.; 20 quart,	85
Pencils. Indelible Label Pencil. Black lead, 5c. each......doz.	50
Picks, Clay or Railroad Picks, handled. Light, 75c.; medium......	85
Pick Mattock. Axe handled......	1.00
Axe Mattock. Axe handled......	1.25
Post Rammer. Cast-iron head......	85
Pruners, Waters' Tree. The best pruner on the market; cuts limbs up to ¾ inch in diameter; with four foot handle, 75c.; six foot, 85c.; eight foot, $1.00; ten foot, $1.15; twelve foot, $1.25. Extra knives......each,	20
"Telegraph" Pruner, attaches to pole of any length, operates with a cord; a spring throws knife back in position......	1.00
The "Victor" Tree Pruner. Very rapid work may be done with this; it has a chisel blade, 2½ inches wide, of the finest steel, attached to a pole twelve feet long that comes in two sections. For low limbs use one section; for tall limbs the two sections screwed together. A long steel guide extends beyond the knife, so you cut the limb at every thrust just where you want to......	2.00
Rakes. (For Lawn Rakes, see page 53.)	
Malleable Iron Garden Rakes, 10 teeth, 25c.; 12 teeth, 30c.; 14 teeth, 35c.; 16 teeth......	40
Steel Garden Rakes, 6 teeth, 30c.; 8 teeth, 35c.; 10 teeth, 40c.; 12 teeth, 45c.; 14 teeth, 50c.; 16 teeth......	55
Short Tooth or Gravel Steel Rakes, 12 teeth, 45c.; 14 teeth, 50c.; 16 teeth......	55
Steel "Bow" Garden Rake. The best rake made; does not break off in the middle. 11 teeth, 50c.; 13 teeth, 55c.; 15 teeth......	60
Wooden Hay Rakes. Best quality, selected wood, three bow. 12 teeth, 30c. each......per doz.	3.00
Old-Fashioned Wood-head Garden Rake, with wrought-iron teeth; strong and durable. A splendid seed-bed rake. 10 teeth in 15½ inch head, 50c.; 12 teeth in 19 inch head......	60
Rubber Plant Sprinklers—	
Straightneck, 6 oz., 45c.; 8 oz., 55c.; 10 oz......	65
Angle neck, 6 oz., 50c.; 8 oz., 65c.; 10 oz......	75
Rubber Putty Bulb, Scollay's. Used in glazing, $1.00; or mailed....	1.10
Saws—Pruning, Lightning Double-edge, 16 inch, 60c.; 18 inch......	70
20 inch......	80
Paragon Double-edge. Thrust on the convex edge, draw on the concave edge, 16 inch, 70c.; 18 inch, 80c.; 20 inch......	90
One-edged Pruning Saw, 16 inch, 50c.; 18 inch, 60c.; 20 inch...	70
California Bow-shaped, 12 inch, 55c.; 14 inch......	65
Little Giant Pruning Hook and Saw Combined. Saw can be removed when desired; attaches to pole of any length......	1.50
Pruning Saw and Chisel Combined...	90
V-Shaped Combination Chisel and Saw Pruner. Handle slips out so that pole of any length may be used......	1.00
Wood Saw, with frame, common pattern......	60
" " **Lightning Tooth, Improved Brace Frame...**	1.00
Saw Buck. Disston's improved pattern...	50
Hand Saws. Finest steel, crosscut or rip. (State which is wanted.) 16 inch, 50c.; 20 inch, 60c.; 24 inch, 70c.; 28 inch......	80
One-Man Crosscut Saw, with supplementary handle. Disston's. Three foot, $2.50; four foot, $3.25; five foot, $4.00; six foot......	5.00
Saw Set, "Taintor's Positive." Will not slip or mar; has but one gauge to set, and any set may be reproduced......	1.00
Scythes—	
Hollow Clipper Grass Scythes. Best quality only, 32 inch, 55c.; 34 inch, 60c.; 36 inch, 65c.; 38 inch, 70c.; 40 inch......	75
Clipper Grain Scythe, 85c.; **Clover Scythe...**	75
Weed and Bramble Scythe, 60c.; **Bush Scythe...**	60
Lawn Scythes. (See page 53.)	
Scythe Snaths or Handles, with patent fastening......	75
Scythe Stones, Imported English Talacre. 25c. each......doz.	2.50
Imitation English Talacre, 15c. each......doz.	1.50
"Red End" flat stone, 5c. each......doz.	50
Scythe Rifles. Best quadruple emery-coated, 10c. each......doz.	1.00
Scissors, Grape Thinning, English. For thinning grapes out of the bunches; best steel. Imported......each,	65
Flower Picking, English. Imported......	75
American Flower and Grape Gathering Scissors. Large and perfect, do not tire the hands, and are made for all-day use......	85
Bouquet Clipping Scissors. Handy and light......	50
Shears—(For Hedge Shears, see page 53.)	
Ladies' Light Pruning, 7 inch......	75
Pruning, Solid Steel, 7½ inch, 86c.; 8½ inch, 90c.; 9½ inch......	1.00
Pruning, malleable, with steel blades, 8 inch......	50
Levin's Improved, medium size, 60c.; full size......	75
English Pruning Shears. No spring. 4 inch, 85c.; 5½ inch......	1.00
" " **Secateurs.** 6 inch, $1.35; 7½ inch......	1.50
Lopping Shears. Imported. No. 1, 20 inch handles, $1.75; No. 2, 24 inch handles, $2.50; No. 3, 28 inch handles, $2.50; No. 4, 32 inch handles......	3.00
American Lopping Shears...	2.00
Ladies' Wood handle Garden Shears...	1.75
Sheep Shears. "True Vermonter." The best quality and shape, especially for sheep shearing. 6 inch blade......	85
Sheep Shears. "Standard Grade," also 5 inch blade......	25
"Curved Handle" Grass Shears. 7 inch blade, finest quality......	1.00
Horse Clipping Shears. Best quality; 8 inch bowed blades......	1.00
Horse Clippers. Best quality......	2.00

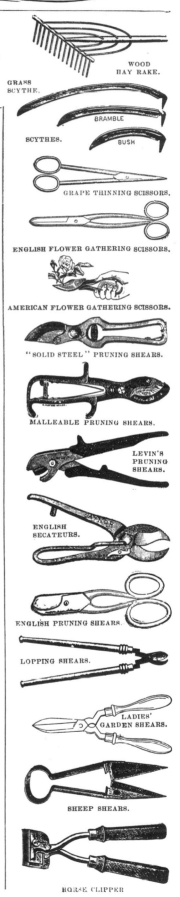

WOOD HAY RAKE.

GRASS SCYTHE.

BRAMBLE

SCYTHES.

BUSH

GRAPE THINNING SCISSORS.

ENGLISH FLOWER GATHERING SCISSORS.

AMERICAN FLOWER GATHERING SCISSORS.

"SOLID STEEL" PRUNING SHEARS.

MALLEABLE PRUNING SHEARS.

LEVIN'S PRUNING SHEARS.

ENGLISH SECATEURS.

ENGLISH PRUNING SHEARS.

LOPPING SHEARS.

LADIES' GARDEN SHEARS.

SHEEP SHEARS.

HORSE CLIPPER.

THERMOMETERS.

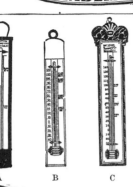
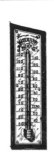

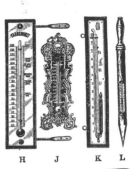

A   B   C   D   E   CLOCK THERMOMETER.   F   G   H   J   K   L

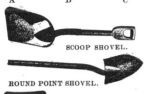

SCOOP SHOVEL.

ROUND POINT SHOVEL.

SQUARE SHOVEL.

WOOD GRAIN SCOOP.

WIRE POTATO SCOOP.

SPADE.

ROUND POINT SPADE.

TREE DIGGING SPADE.

DITCHING SPADE.

POST HOLE SPADE.

FLAT DRAIN SPADE.

POST HOLE SPOON.

SWEAT SCRAPER.

GARDENER'S TAPE LINE.

TREE SCRAPER.

		Price.
Shovels—		
Square. No. 2, D-handle; standard quality, steel, 65c. each.....doz.		7.50
" " No. 2, Long " " " " 65c. " doz.		7.50
Round Point. No. 2, D-handle; standard quality, steel, 65c....doz.		7.50
" " No. 2, Long " " " " 65c....doz.		7.50
Ames' Square. No. 2, D-handle; extra quality, steel.............each,		1.25
" " No. 2, Long " " " " ......each,		1.25
" Round Point. No. 2, D-handle; extra quality, steel....each,		1.25
" " No. 2, Long " " " " ....each,		1.25
Scoop Shovels. No. 2 (10½ x 14½ inches), D-handle, steel.....each,		85
" " No. 3 (11¼ x 15 inches), D-handle, steel.....each,		90
Grain Scoop. Wide mouth (13½ x 16 inches), D-handle, steel.......		1.00
Wood Grain Scoop. All wood ....each,		1.00
Wire Potato Scoop. For handling roots, etc.; dirt sifts out, each,		1.50
Boys' Shovels and Boys' Spades..each,		50
Spades—		
Steel, Standard quality. No. 2, D-handle, 65c. each.... ....doz.		7.50
" " " No. 2, Long handle, 65c. each..........doz.		7.50
Ames', Extra quality. No. 2, D-handle..................................each,		1.25
" " " No. 2, Long handle......................each,		1.25
Round Point Spade..each,		1.25
Nursery Spades. Tapered, extra heavy; D-handle, strapped........		1.50
Tree Digging Spade. Tapered, large, heavy and extra strong, each,		3.00
Drain Spades. D-handle, with foot clasp, 18 inch.....................each,		1.50
Concave Post Hole Spade. D-handle, 16 inch, $1.50; 18 inch.....		1.75
Ditching Spade. D-handle, 16 inch, $1.50; 18 inch, $1.75, 20 inch,		2.00
Post Hole Spoon, 10 x 8¼ inches		1.75
Flat Drain Spade, square point...		1.50
Sweat Scraper. For removing sweat, dandruff, etc., from horses, each,		60
Tape Line. Pressed leather case, ½ inch Holland tape. 25 foot, 50c.;		
50 foot, 75c.; 75 foot, $1.00; 100 foot..		1.25
Thermometers—(For Dairy, see page 45.) (For Incubator, see page 57.)		
Common Japanned Tin Case. (Fig, A.) (Common grade.)		
7 inch, 15c.; 8 inch, 20c.; 10 inch, 25c.; 12 inch		30
Household Wood Back, black enameled (Common grade), 8 inch,		25
" " " " white " (Fig. B) " 8 "		25
Household Fancy Carved Oak (Fig. C) (Medium grade), 8 inch,		
$1.00; 10 inch		1.25
Household Wood Back, metal scale (Medium grade), 8 inch, 45c.;		
10 inch		55
Household "Distance Reading" (Fig. D), magnifying tubes,		
bold figures, metal scale, natural wood back (Standard or best		
grade), 12 inch..		1.50
Household Porcelain Scale (Fig. E), oak back, magnifying		
tubes, cylindrical bulb, 8 inch, $1.00; 10 inch		1.50
"Radial Scale" Suspending Thermometer (Fig. G), with		
chain to hang from chandelier or bracket, furnished either in		
polished brass or nickel plate with black oxidized scale............		2.00
Fancy Metal Standing Thermometer (Fig. J), silver panel		
and black oxidized scale, 8½ x 3 inches, $1.25 each; 11½ x 4		
inches ....each,		1.50
Plate Glass Window Thermometer (Fig. H), with arms for		
attaching, bevel edge, white enamel face, 8 inch, $1.25; 10 inch...		1.50
Siexe's Heat and Cold Self-Registering Thermometer, 8		
inch, $2.50 each; 10 inch (Fig. F)..each,		3.00
Minimum Registering Thermometer, 8 inch wood case (Medi-		
um grade). (Fig. K)...		85
Maximum Registering Thermometer, 8 inch wood case (Medi-		
um grade). (Fig. K)...		90
Hot-Bed or Mushroom-Bed Thermometer, pointed brass bot-		
tom for plunging. (Fig. L)..		2.00
Clock Thermometer, 9 inch dial, metal case; very sensitive and		
quickly read...		2.00
Tree Scraper, handled		50
Transplanter for tomatoes and large plants...............................		50
Transplanting Trowel. Solid steel; largely used by gardeners in		
putting out cabbage plants, etc..		30
Trowels—Ordinary, 5 inch, 10c.; 6 inch, 15c.; 7 inch..............		20
Solid Steel, 5 inch, 40c.; 6 inch, 45c.; 7 inch.....................		50
Cleves's Angle Trowel. Small size, 20c.; large size		30
Watering Pots. Galvanized iron, 6 quart, 75c.; 8 quart, 90c.; 10		
quart, $1.00; 12 quart. $1.25; 16 quart....................................		1.50
French or Oval Shape Watering Pots. 6 quart, $1.50; 8 quart......		1.75
"Common Sense" Watering Pots, 2 quart, 75c.; 4 quart, $1.00;		
6 quart..		1.25
Wotherspoon's Galvanized Watering Pots, with fittings and two		
brass roses, one fine and one coarse, with each pot. 6 quart, round,		
$1.75; 8 quart, round, $2.00; 10 quart, round, $2.25; 12 quart,		
round, $2.50; 16 quart, round		3.00
Weeders—Excelsior............... 10 **3-cornered**..................		20
Hazeltine's .... 20 Jennings'..........................		15
Eureka Hand Weeder, the best little tool for loosening the soil		
around plants that we know of; thin forged steel fingers............		25
Lang's Weeder ....each,		20
Wrench, Agricultural Screw Wrench. 8 inch, opens 1¼ inch, 60c.;		
10 inch, opens 1¾ inch, 75c.; 12 inch, opens 2⅛ inch..................		1.00

STEEL TROWEL.

ANGLE TROWEL.

TRANSPLANTER, FOR LARGE PLANTS.

TRANSPLANTING TROWEL.

EXCELSIOR WEEDER.

HAZELTINE'S WEEDER.

ONION WEEDER.

JENNING'S WEEDER.

EUREKA WEEDER.

LANG'S WEEDER.

AGRICULTURAL SCREW WRENCH.

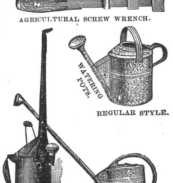

WATERING POTS.

REGULAR STYLE.

COMMON SENSE.   "FRENCH."

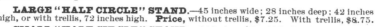

REVOLVING ADJUSTABLE STAND.

HENDERSON'S SUPERIOR

WIRE PLANT STAND PACKED FOR SHIPMENT.

SMALL SINGLE SHELF STAND.

WIRE PLANT STANDS.

These "knock down" for shipment, and consequently reach their destination in perfect condition. They are very strong and steady and are handsomely finished in green enamel and gold.

LARGE "HALF CIRCLE" STAND.—45 inches wide; 28 inches deep; 42 inches high, or with trellis, 72 inches high. **Price,** without trellis, $7.25. With trellis, $8.75.

SMALL "HALF CIRCLE" STAND.—42 inches wide; 26 inches deep; 42 inches high, or with trellis, 67 inches high. **Price,** without trellis, $6.25. With trellis, $7.75.

THREE SHELF SQUARE STAND.—41 inches long; 25 inches deep; 42 inches high, or with trellis, 75 inches high. **Price,** without trellis, $7.25. With trellis, $8.75.

TWO SHELF SQUARE STAND.—34 inches long; 17 inches deep; 32 inches high, or with trellis, 60 inches high. **Price,** without trellis, $4.50. With trellis, $6.00.

SMALL SHELF SQUARE STAND.—28 inches long; 10 inches deep; 22 inches high. **Price,** $3.50.

SQUARE STAND, WITHOUT TRELLIS.

HALF CIRCLE STAND, WITH TRELLIS.

REVOLVING ADJUSTABLE PLANT STAND.

No. 2.—Has two tiers of brackets and holds 17 pots. Diameter, 26 inches; height, 4 feet; weight, about 25 lbs. **Price,** $5.00.
No. 3.—(See cut.) Has three tiers of brackets and holds 23 pots. Diameter, 32 inches; height, 5 feet; Weight, 35 lbs. **Price,** $6.50.

PORTABLE OIL HEATERS.
Heat by Radiation.

No Smell !   No Dirt !   No Gas !

Just what is wanted for small conservatories, window gardens, bathrooms, small bedrooms, etc.; invaluable for protecting your plants on cold nights.

They are made of brass, nickel-plated, and have Russia iron cylinders. The combustion is perfect, therefore absolutely free from the offensive odor and smoke.

The Mica Lining allows a pleasant light to shine through the open work of the cylinder.

No. 10 Heater.—Weighs 4¼ lbs., stands 23 inches high; circumference of drum, 18 inches; holds two quarts of oil; will burn nine hours, and will heat an 8 x 10 room. **Price,** $3.50.

No. 44 Heater.—Weighs 8 lbs., stands 32 inches high; circumference of drum, 25 inches; holds four quarts of oil; will heat a room 15 x 20 feet; will burn from eighteen to twenty hours. **Price,** $6.00.

WINDOW BRACKETS FOR PLANTS.

4-POT BRACKET.

Highly finished and supplied complete with screws.

	Each
1 pot	$0.25
2 "	.50
3 "	.75
4 "	1.00

Hanging Basket and Bird Cage Hook, 15cts. each.

No. 44 PORTABLE OIL HEATER.

Indurated Fibre Waterproof Saucers.

Are not porous, and will protect woodwork, tables, etc., on which plants are to stand. Not breakable.

4 in....ea.	6c.	9 in.....ea.	9c.
5 " "	7c.	10 " "	10c.
6 " "	7c.	11 " "	11c.
7 " "	8c.	12 " "	12c.
8 " "	8c.	13 " "	15c.
14 in....ea. 18c.			

CUT FLOWER VASE.

INDURATED FIBRE WARE

White Enamelled Indurated Fibre CUT FLOWER VASES. Do Not Break.

Depth	Diam.		Price	Depth	Diam.		Price
4½ in.	x 3 in.		30c.	12 in.	x 4 in.		50c.
6 "	x 4 "		35c.	15 "	x 4½ "		60c.
9 "	x 4½ "		45c.	13 "	x 8 "		75c.

ROLLING STANDS FOR HEAVY PLANTS.

A very useful waterproof saucer arrangement on ball-bearing casters for turning or moving heavy plants, and preventing injury to carpets from drip or dampness.

Diameter	Will Carry	On Casters	Price
13 in.	11 in. Pot	3	$0.75
18 "	16 " Tub	4	1.00
20 "	18 " "	5	1.25
22 "	20 " "	6	1.50

"STANDARD" FLOWER POTS.

☞ No order filled for less than $2.00 worth.   We pack carefully, but will make no allowance for breakage.

Breakage is not one-half as great as in other pots, the deep rim protecting them. The foot keeps the pot up from bench, so that it is impossible for hole to become clogged. The concave bottom and large hole insure perfect drainage.

Standard Flower Pots.

	Per doz.	Per 100
2 inch	$0.12	$0.75
2½ "	.15	.90
3 "	.20	1.25
4 "	.35	2.00
5 "	.45	2.75
6 "	.75	4.25
7 "	1.25	6.50
8 "	1.50	9.00
10 "	3.00	18.00
12 "	4.00	27.00

Common Flower Pot Saucers.

	Per doz.
4 inch	$0.25
5 "	.30
6 "	.35
7 "	.40
8 "	.50
9 "	.85
10 "	1.00
11 "	1.25
12 "	1.50
14 "	1.75

Round Seed or Lily Pans.

H'g't Width	Per doz.
4 x 8 inches	$1.50
5 x 10 "	1.75
6 x 12 "	2.00
7 x 14 "	2.25
8 x 16 "	2.50
9 x 18 "	3.75

Square Seed Pans.

	Per doz.
6 x 6 inches	$2.00
8 x 8 "	2.50
10 x 10 "	3.00
12 x 12 "	3.50

Hoop Flower Pots.

	Each	Per doz.
8 inch	$0.30	$2.75
9 "	.40	4.00
10 "	.55	5.50
12 "	.70	7.00
14 "	1.00	10.00
16 "	1.60	16.00
18 "	2.50	25.00
20 "	3.80	38.00
24 "	5.50	56.00

Terra Cotta Hanging Pots.

Hanging Log, 7 in.	30c.
" 9 "	45c.
" Shell, 11 "	90c.
" Round } 7 "	30c.
" Rustic }	
" Round } 9 "	40c.
" Rustic }	
Brass Chains, extra, ea.	15c.

PAT'D JUNE 30.74.

"Rocking Table" Apple Parer.

With push-off so arranged that parings and juice cannot fall upon it; pares close to the fork and does good work; the improved clamping device will not jam the table. Pr. 75c.

Apple Parer, Corer and Slicer.

The parings and juice fall clear of the working parts; strong, durable, and does good work; it can be used to pare only without coring and slicing, if desired. Price, $1.50.

Potato Parer.

Takes off a thinner paring from potatoes of all shapes and kinds, and does it much better than can be done by hand; it even goes into and cleans out the eyes; it will also pare quinces and pears as well as potatoes. Price, 75c.

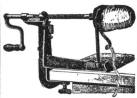

Cherry Stoners.

"The Rapid" adapted for rapid work; adjusts to different sizes of stones. Price, 75c.

"The Perfect" works more slowly than the above, with the least possible cutting or disfiguring, an advantage in preserving. $1.00.

Grape and Raisin Seeder.

Seeds grapes or raisins either wet or dry, though does better if the raisins receive the usual washing, then every seed will be removed without waste. Family size, will seed a pound in 5 minutes, $1.00; hotel size, will seed a pound in 1 minute, $2.50.

Fruit, Wine and Jelly Press.

Superior, simple, easily operated, saves all re-handling, as it extracts the juice and ejects the skins and seeds in 1 operation; it can be used for many purposes, including the making of wine, jellies, fruit butters from grapes, currants, quinces, pineapples, tomato catsup, etc. Family size, capacity 1 qt., $3.00; large size, 3 qts., $15.00.

Sausage Stuffer and Lard or Fruit Press.

For stuffing sausages the tin strainer and bottom plate are removed; for pressing fruit or lard they must be replaced; the pressure will remain without holding the crank; 2 plungers are furnished, 1 for stuffing and 1 for pressing. 2-qt. size, $4.00; 4-qt., $6.00; 8-qt., $7.50.

Meat Chopper.

The meat is cut perfectly, not torn; adapted for making sausage, mince meat, hash, hamburger steak, croquettes, chicken and lobster salads, chopping peppers, cocoanuts, horseradish, meat for beef tea, corn for fritters, etc. Small size, 1 pound a minute, $2.00; large size, 2 pounds a minute, $3.00.

THE EUREKA STEAM COOKER.

VEGETABLES, etc., cooked by steam are greatly superior in flavor, nutritiousness and dryness to those cooked by boiling. Vegetables are composed of from 70 to 90 per cent. of water, and when immersed in more water and boiled they have fully one-third of their food values washed out, and consequently are frequently tasteless and watery. If you enjoy nice, dry cooked vegetables, that retain their rich natural flavor, cook them by steam. Steam cooking is practiced largely by the French and by a few in this country, none of whom could be induced to go back to boiled vegetables again; therefore, there is nothing new or original about it, but this recently invented Eureka Steam Cooker embodies such improvements and new features that we feel impelled to draw our customers' attention to it, even if it is "a little out of our line." It is a series of vessels, one fitting on top of the other, as shown in the illustration, thus allowing a number of things to be cooked at one time and on one hole of any stove or range, or even on an oil or gas stove. The lowest vessel is the boiler, which is filled with pure water **only**. This forms steam, which passes up, over and into the other vessels, each independently of the other. This steam cooks even more thoroughly and just as quickly as boiling or roasting. As the steam condenses it drips from each pan into a drainage tube, and not from one vessel to the other. This drainage tube carries all drip into the condenser, which is also used for making soup or soup stock, just over the boiler, thus the water in the boiler is always pure. There is no mixing of flavors in the separated vessels and no cooking odor in the house, for the only steam that escapes is from the fresh water. **The Eureka Steam Cooker** is made for use, of tin, twice as heavy as ordinary tinware; boiler is copper bottomed. Every cooker is guaranteed to do as represented. Full directions with each.

The Eureka Steam Cooker will cook anything that can be boiled, baked or roasted, vegetables of all kinds, fruits, cereal foods, meats, eggs, oysters, soups, puddings, sauces, etc. A whole dinner may be cooked at one time, and everything will be delicious. A good cook can cook better with it, and a poor cook cannot spoil the meal if she tries. No watching, no boiling over, no scorching, no waste of fuel or time. A meal may be kept for hours with this cooker without spoiling. Give everything the usual length of time. Cooking receipts can be followed with this cooker just the same.

For Canning Fruits. The Eureka Cooker is very valuable; fruits retain their natural shape and color. No stirring is required, which mashes and breaks the fruit. Simply put the raw fruit in the cans, either glass or tin, place in the cooker and steam 25 to 35 minutes and seal immediately. Canned vegetables are just as easily and perfectly done.

For Sterilizing Milk. To keep it, or for babies' or invalids' food, this cooker is perfectly adapted. For this purpose we supply a rack and graduation bottles for $1.00 extra.

For Distilling Water. To purify it for drinking or other purposes, it will be found useful.

Family Slaw Cutter and Vegetable Slicer.

Slices, thin or thick, cabbage, potatoes, onions, cucumbers, citron, etc.; the vegetable is simply held on the platform and pushed against the revolving cylinder, which contains 3 knives. Price, $3.00.

Large Vegetable and Cabbage Slicer.

For slicing, thin or thick, vegetables and fruits, such as cabbage for kraut or slaw, potatoes, beets, turnips, carrots, cucumbers, egg plants, pumpkins, squash, onion, citron, melon rind, pineapples, apples for drying, etc.; easily operated. Price, $5.00.

Vegetable Grater.

For grating horseradish, cocoanuts, apples, corn for fritters, crackers, cheese, etc.; the material to be grated is placed on the platform and simply pushed against the revolving cylinder. Price, $3.00.

Prices and Sizes of Eureka Steam Cookers.	Number People.	Vessels.	Quarts each.	Diameter.	Height.	Price.
Hotel Size	20 to 30	7	10	12 in.	37 in.	$10.00
Boarding House Size	12 to 20	6	8	10 "	31 "	6.50
Extra Large Family Size	8 to 12	5	6	9 "	26 "	4.50
Large Family Size	6 to 8	4	6	9 "	21 "	3.75
Medium Family Size	4 to 6	4	4	8 "	20 "	3.25
Small Family Size	3 to 4	4	2½	7 "	15 "	2.25
Students, etc., Size	2 to 3	3	2½	7 "	12 "	1.75

White Mountain Ice Cream Freezer.

Freezes cream quick and smooth. 2-qt., $2.00; 3-qt., $2.50; 4-qt., $3.00; 6-qt., $3.75; 8-qt., $5.00.

Family Scales.

Durable, finely finished, warranted accurate. Capacity, 12 lbs., $2.75; 24 lbs., $3.50; 48 lbs., $5.00.

The "Acme" Corn Popper.

A patent lever closes, opens and holds open the lid; the lever is always cool; these poppers are full size, and much stronger and better than the ordinary. Prices, 1-quart size, 20c. each; 1½-quart size, 25c.; 2-quart size, 30c.

Family Pop Corn Sheller.

The sheller is held in one hand, the ear of corn in the other; it shells easily, quickly and without scattering. Price, 25c. each, or mailed, 30c.

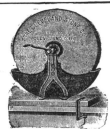

Family Grindstone.

Mounted, cast iron basin, and clamp to fasten to table. 6-inch (diameter), 75c.; 8-inch, $1.00; 10-inch, $1.25; 12-inch, $1.50.

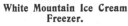

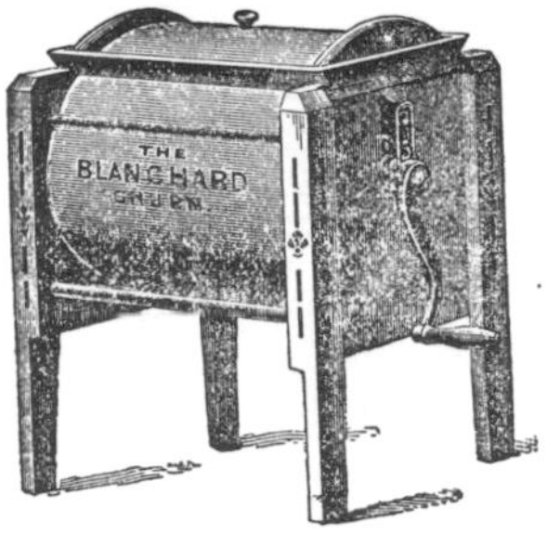

THE BLANCHARD CHURN

Has retained its reputation for over twenty years as being the best revolving dash Churn made for ease of operation and superior butter-making qualities.

No.		No.	
3. Ch'ns 2 gal.	$6.00	6. Ch'rns 12 gal.	$10 00
4. " 4 "	7.00	7. " 16 "	12.00
5. " 8 "	8 00		

Factory sizes quoted on application.

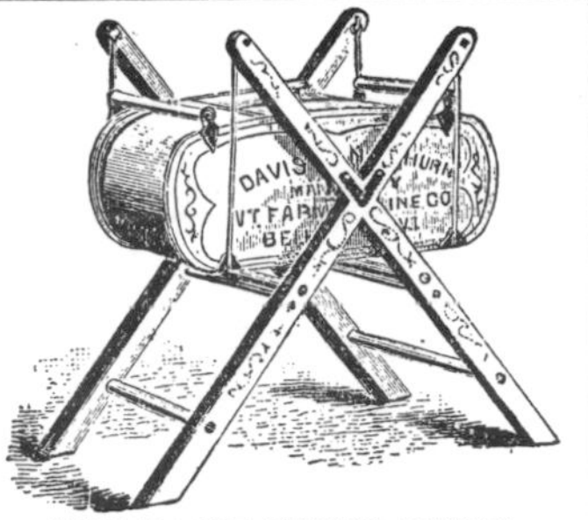

THE DAVIS SWING CHURN

Is constructed on correct scientific principles. A swinging motion keeps the entire mass in motion, and quickly brings the butter by the particles of cream coming in contact with each other. Easily cleaned because there is no dasher.

No.1. Ch'ns 3 gal.	$7.	No.4. Ch'ns 12 gal.	$12.
No.2. " 5 "	8.	No.5. " 15 "	15.
No.3. " 8 "	10.	No 6. " 20 "	18.

Prices of larger sizes on application.

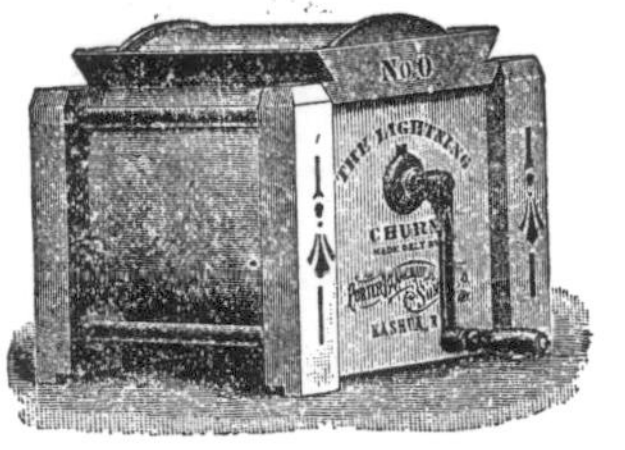

LIGHTNING CHURN.

This supersedes the old revolving dasher cylinder Churn, being an improvement over it in many ways, more substantial and at but little extra cost.

No. 0.	Churns 2 gallons	$3.00
No. 1.	" 3 "	3.50
No. 2.	" 4 "	4.00
No. 3.	" 5 "	4.50

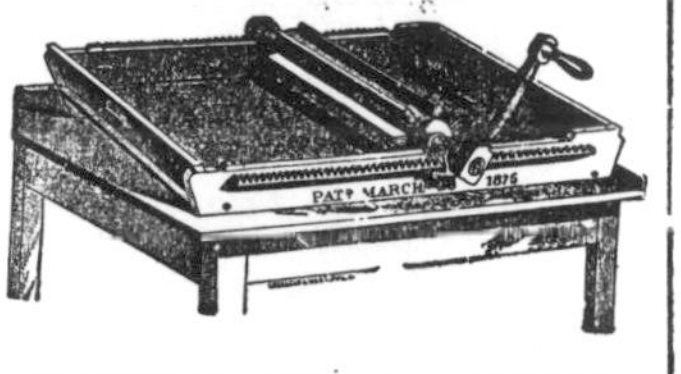

IMPROVED BUTTER WORKER.

Choose a size that will work the largest quantity you will be likely to work, for two lbs. can be worked in any of the machines as well as the maximum quantity.

10-lb. Butter Worker	$6.00
20-lb. "	7.00
30-lb. "	8.00
50-lb. "	10.00

Larger sizes quoted on application.

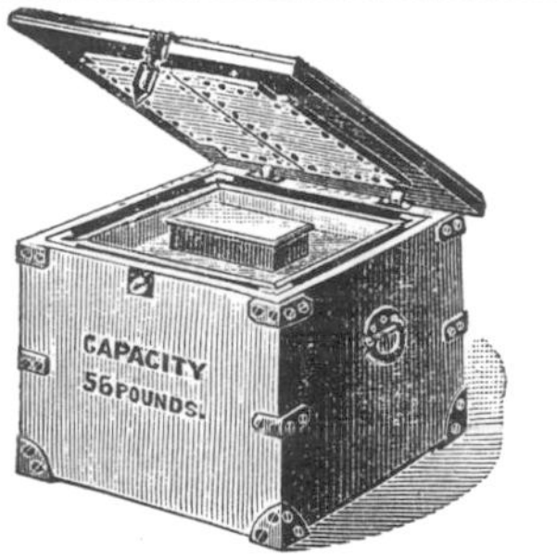

Improved Butter Shipping Box.

These boxes are iron bound at the edges and capped at the corners. An ice box is fitted in centre, which can be removed at will and the space utilized.

CAPACITY.

12 1-lb. Prints..	$3.50	56 1-lb. Prints.	$5.00
20 1-lb. "	3. 5	80 1-lb. "	5.50
30 1-lb. "	4.25	96 1-lb. "	5.75
40 1-lb. "	4.75	120 1-lb. "	6.00

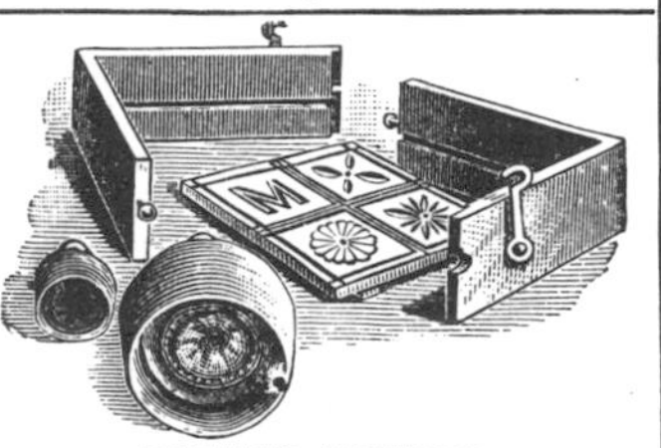

BUTTER MOULDS.

BLANCHARD'S UNLOCKING MOULDS.

½-lb. size, marking the 2 quarters	$1.15
1-lb. " " 4 "	1.35
2-lb. " " 8 "	1.85

ROUND MOULDS.

½-oz. "Individuals"	20c. each;	$2.00 doz.
1-oz. "	25c. "	2.50 "
¼-lb. Mould	30c. "	3.00 "
½-lb. "	40c. "	4.00 "
1-lb. "	50c. "	5.00 "

The "Gravity" Cream Separator.

RAISES CREAM IN TWO HOURS

WITHOUT ICE OR POWER.

SO REASONABLE.. IN PRICE that no one who makes butter can afford to be without it!

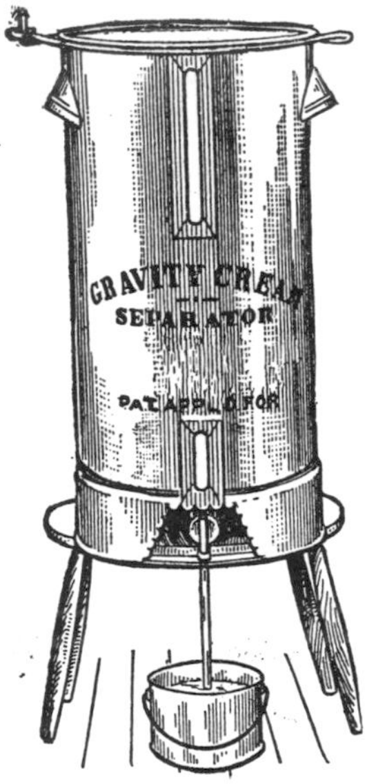

The "Gravity" Cream Separator is a new device for separating cream from milk where there is no objection to the "skim milk" being diluted with water—for instance, when fed to hogs, etc. It raises the cream in two hours, and is the cheapest and best method for butter makers to handle their milk.

There is a property in milk called viscosity. It is caused by the solids in the milk, aside from the butter fat. It is a sort of stickiness that retards the raising of the cream. By proper dilution with water we so reduce the viscosity that the cream will raise in our Separator in two hours, and on the same principle of raising cream quickly it also raises it more thoroughly than with the creamery, the loss being only about three-tenths of one per cent. Those that keep only one or two cows, as well as the large dairymen, can have the advantage of a Separator at a small cost, compared with the centrifugal Separator, or creamer.

You will make more and better butter from the same number of cows. It reduces the cost of handling milk very much. It does away with the handling, storing and cost of ice, and there is no machinery to keep in order. It saves work, and makes life easier for the women folks, and is easy to wash and keep clean. It runs itself, requiring no attention, and is well made and durable.

Full Directions are Furnished with each Separator.

PRICES.

No. 1.	For 1 to 4 cows;	capacity 100 lbs. milk per day		$6.00
No. 2.	" 5 to 7 "	" 200 " "		9.00
No. 3.	" 8 to 15 "	" 400 " "		11.00

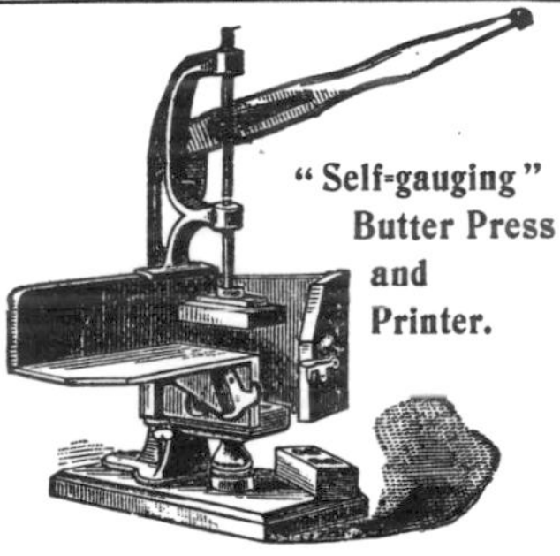

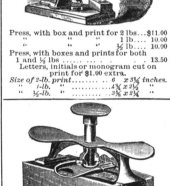

"Self-gauging" Butter Press and Printer.

Press, with box and print for 2 lbs...	$11.00
" " " 1 lb....	10.00
" " " ½ lb....	10.00
Press, with boxes and prints for both 1 and ½ lbs	13.50

Letters, initials or monogram cut on print for $1.00 extra.

Size of 2-lb. print .... 6 x 3¼ inches.
" 1-lb. " .... 4¾ x 2½ "
" ½-lb. " .... 3⅝ x 2¼ "

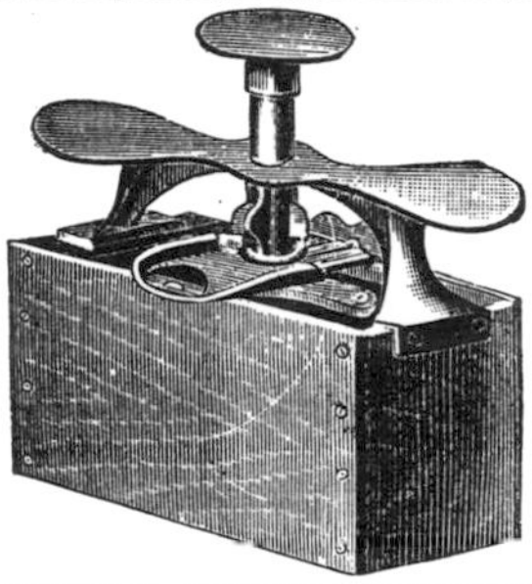

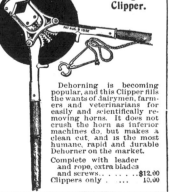

EUREKA BUTTER PRINTER.

This improved Butter Mould and stamp is self-gauging, giving accurate weight when correctly adjusted.

Moulds and prints pounds 4⅝ x 2½ x 2¾ in.
" half " 4⅝x2½x1 3-16 in.

Price, with uncarved print block....	$3.50
" with design or letter in "	4.50

SAFETY MILKING INSTRUMENT.

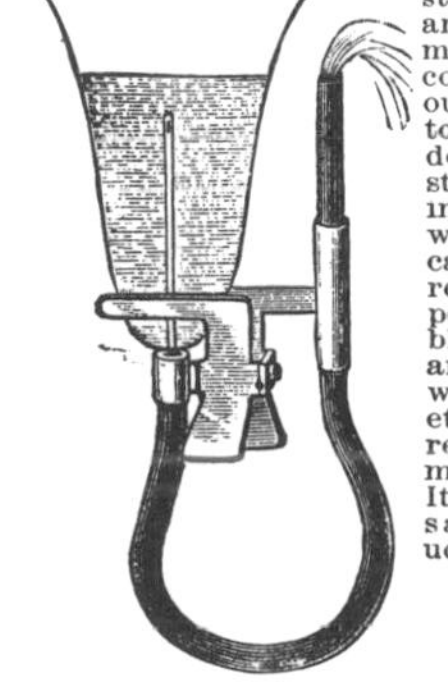

The only instrument guaranteed to draw milk from a cow's bag without allowing air to enter, as it does with other styles of milking tubes, and which often causes serious results. It is especially valuable when cows are troubled with sore teats, etc., and are restless when milked by hand. Its use often saves a cow's udder.

$1.00 each.

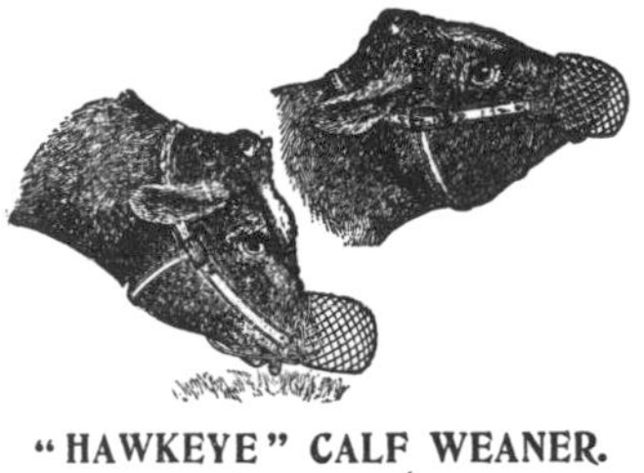

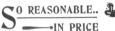

"HAWKEYE" CALF WEANER.

Prevents calves from sucking, and yet allows them to eat and drink from the ground. It does not fasten to the nose like other Weaners, but is strapped on. There are no prongs to do injury; made of wire, tinned to prevent rusting.

PRICES, COMPLETE WITH STRAPS.

For calves	$0.60 each;	$6.00 per doz.
For yearlings	.80 "	8.00 "
For cows	1.00 "	10.00 "

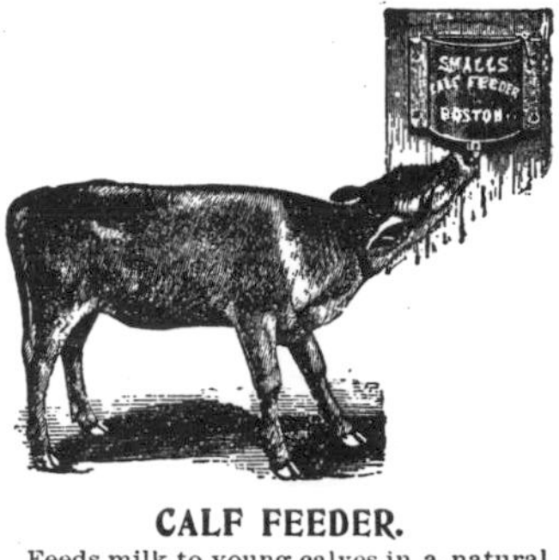

CALF FEEDER.

Feeds milk to young calves in a natural manner before they are old enough to eat and drink otherwise.

Price	$2.50
Extra nipples....	25c. each, $2.50 per doz.

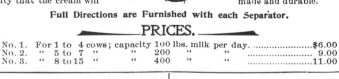

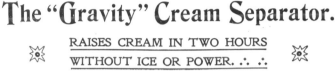

Dehorning Clipper.

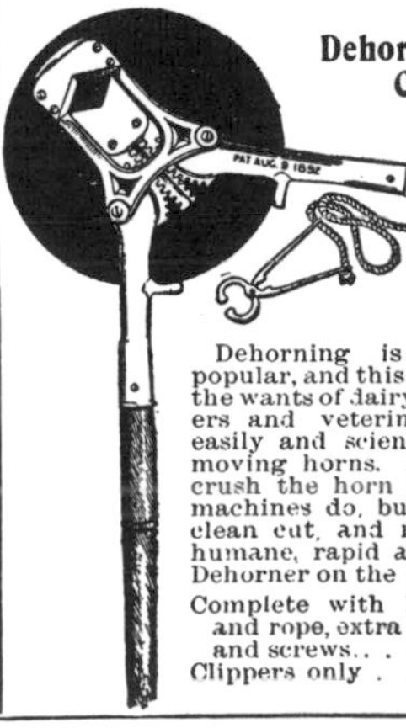

Dehorning is becoming popular, and this Clipper fills the wants of dairymen, farmers and veterinarians for easily and scientifically removing horns. It does not crush the horn as inferior machines do, but makes a clean cut, and is the most humane, rapid and durable Dehorner on the market.

Complete with leader and rope, extra blades and screws	$12.00
Clippers only	10.00

COOLEY MILK STRAINER.

Strains rapidly, will not clog, and strains better than any other, as finer wire cloth is used than could be done with a flat strainer. Can be used on small pans or cans.

Price, with rest, $1.75.

"Elevator" Style. COOLEY CREAMERS. "Cabinet" Style.

These are acknowledged the best apparatus for separating cream from milk by the gravity process. The "Elevator" Style is the most popular, as the cans can be raised and lowered into the water by the elevator. The "Cabinet" Style is preferable when milk is to be drawn off, from under the cream, before skimming. Each can measures 19 inches high by 8½ inches in diameter, and holds 18 quarts or the milk of from 2 to 3 cows.

Prices, "Elevator" Style. INCLUDING CANS.		Prices, "Cabinet" Style. INCLUDING CANS.	
2 can size......$32.00	8 can size......$68.00	1 can size......$20.00	4 can size......$40.00
4 " " 45.00	10 " " 79.00	2 " " 27.00	6 " " 52.00
6 " " 57.00	12 " " 93.00	3 " " 33.00	8 " " 64.00

(*Larger sizes quoted on application.*)

"DE LAVEL"
Centrifugal Cream Separator.

Rapidly and thoroughly separates cream from milk by centrifugal force; few parts; simple, and easily kept clean.

Prices.
"Strap," Humming Bird style, (see cut)-$50.00
"Crank," - - - - - - - 65.00
Old style Baby No. 1 - - - - - 50.00
Improved " - - - - - -100.00
Improved Baby No. 2 - - - - -125.00
" No. 3 - - - - -200.00
Improved Dairy Turbine - - - -225.00

(*Illustrations and descriptions mailed on application.*)

STAR MILK COOLER.

This apparatus is designed for cooling with water, milk and cream, automatically and quickly, without the use of ice, at any season of the year. They are constructed of copper sheets and brass castings, heavily tinned, and have **surface enough** to do the work claimed in **hot weather.** There is, of course, no rust out to them, and they will last a lifetime. They are guaranteed to be as represented in every way.

No. 0, cools 50 gals. per hour........$15.00
" 1, " 60 " " 25.00
" 2, " 90 " " 27.50
" 3, " 120 " " 40.00

Syphons, for producing a continuous flow of water from spring or other water supply below level of Cooler, extra, $5.00.

The "Model" Combined Milk Cooler and Aerator
DOES WHAT NO OTHER COOLER WILL.

Rapidly and thoroughly cools and at the same time aerates the milk perfectly, removing all odors.

It *is simple in construction;* any child old enough to milk can use it. The milk is simply poured into the receiving tank as it is milked, so that when through milking, your milk is cooled and aerated ready to ship or deliver to customers or set for cream, as the cream commences to raise at once and is far riper and better than when set the old way.

It can be used either with running water or filled from the well or spring or with ice. The milk will be cooled to within a few degrees of the temperature of the water used in cooling. Once filling with ice will cool 400 to 500 quarts down to a temperature of 50 degrees.

It is a complete deodorizer, for when the milk passes down over the cooling surface, which is, of course, very cold, it causes steam to arise, carrying off any odor in the milk caused by the cows eating roots, cabbage, or weeds that taint milk. Most dairymen know that if cows breathe impure air, eat unclean or flavored food, or drink impure or tainted water, the milk will be affected and have the same taint. The use of the deodorizer removes this taint, rendering the milk at once cool and sweet, and will keep sweet longer than when treated in any

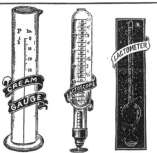

other way. It is sweeter and better for the milk room, and may be carried to the factory in close cans without danger of taint.

DESCRIPTION.—This apparatus spreads the milk over a large surface for cooling, at the same time exposing it to the contact and influence of the air. In the bottom of the Receiver is a double circle of fine holes; the milk runs through these holes and spreads over the large flaring surface, cooled by the water inside, while the air carries off all odors, rendering the milk sweet and cool. When it reaches the gutter it is held in contact with the Cooler at its coldest point while flowing around to milk outlet. It is as easily cleaned as a milk pan.

TO OPERATE.—Set as low as possible where there is plenty of pure, fresh air; if the wind blows, the better. It can be used in the strongest wind without a drop being blown away. Use very cold water.

Use a cloth or metal strainer over the top of Milk Receiver.

Improved Oval Strainers of finest brass wire cloth, capacity 15 square inches, furnished for any size at $1.00 extra. Plungers for use in still water go with each.

PRICES.
No. 1, for 1 to 6 Cows..........$6.00 | No. 3, for 25 to 40 Cows..........$8.00
No. 2, " 10 to 15 " .... 7.00 | No. 4, " 50 to 100 " 10.00
(*Special sizes for cooling cream from Separators quoted on application.*)

The U. S. Improved Centrifugal
CREAM SEPARATOR.

All latest improvements; enclosed safety gears; have attachments for either hand or power, as ordered. The construction of the interior is not complicated, has few parts to clean, and does the most thorough skimming or separating of cream from the milk.

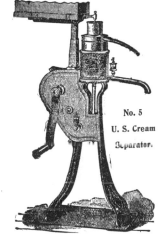

No. 5
U. S. Cream
Separator.

No.	Capacity per hour.	Price.
7, Midget......	200 lbs.	$75.00
6, High frame......	250 "	100.00
5, "	350 "	125.00
3, "	650 "	200.00
3, Low "	600 "	200.00
Power Ratchet, pulley attach'ts, extra		2.50
" " and loose pulley " "		5.00

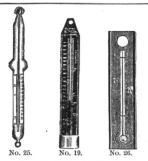

THE "PERFECT"
MILK PAIL AND SEAT.

It cannot be overturned by the cow, as the milker sits on it; the funnel and the pail are connected by a detachable, easily cleaned tube, in which is fixed a strainer so that no dirt can get into the milk.

Price, $2.00.

No. 25. No. 19. No. 26.
DAIRY THERMOMETERS.

No. 25-A. All Glass Floating. Floats upright so whole scale is in view; hand graduated magnifying tube............50c. each.
No. 25B. Same as above; with red spirit tube ....60c. each.
No. 19. Flange Dairy Thermometer. Brass scale and sliding guard, 7 in., 50c. each; 8 in., 60c. each; 10 in., 70c. each.
No. 26. Churn Thermometer. Zinc, 5 in. flanged zinc scale; tested at 62° for churning............15c. each, $1.50 doz.

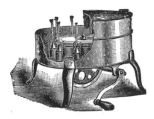

Cream Gauges. Graduated 0-30 for determining percentage of cream in milk. Size, 12x2 ins., 85c.; size, 10x1½ ins., 75c.; size, 9x¾ ins., 50c.; set of 4 in tin stand, $2.00.
Lactoscope. Prof. Feser's simple and direct method of determining amount of cream in milk. Without Foot rest, $3.00; with Foot rest, $4.00.
Lactometer. Indicates adulterated milk and per cent of water with milk at temperature of 60°. **Ordinary,** in wood case, 50c.; N. Y. Board of Health Standard, 75c., or combined with thermometer, $1.75.

BABCOCK MILK TESTER.

The best Tester on the market. With each machine is included full directions for operating. Standard size milk bottles, pipette, acid enough for 50 to 100 tests.

4 bottle tester..........complete, $8.00
6 " " .... " 9.00
8 " " .... " 10.00
10 " " .... " 12.00

THE HENDERSON HIGH WHEEL "BALL-BEARING" Hand Lawn Mower.

INVENTION'S RIPEST FRUIT! HAS BEARINGS LIKE A BICYCLE! THE LIGHTEST WORKING MOWER MADE!

A 24-inch cut pushes as easily as an ordinary Mower of only 16-inch cut.......

A lawn may now be mown in one-half the time required with old style narrow-cut Mowers.

The Henderson Ball-bearing Lawn Mower is superior to all other Mowers in the following respects: The axles of the revolving cutter, like the axles of a bicycle, work in "**ball-bearing**" journal boxes or cups, which reduces the friction enormously. These parts are made of the finest case-hardened steel, and will last for years, though we have provided against wear should there be any—so that any one, by simply loosening one screw and tightening another, can force the cup and balls higher up the cone-shaped ends of the revolving shaft, thereby taking up all possible wear. This adjustment is so simple, yet positive, that it can be set to a hair. Our Mower is also built scientifically correct in other respects, resulting in a phenomenally light running Mower. A 24-inch cut cuts the grass as easily as a 16-inch cut old style Mower. Your gardener will not now complain about your Mower being "too large and a man killer," if you supply him with a "Henderson Ball-bearing" Mower, even of wide cut. On the contrary, the gardeners to whom we sent these Mowers last summer for testing, were enthusiastic about them. As one gardener expresses it: "*I can cut my lawn in four hours and not be half as tired as I used to be with my old Mower when I took a day for it.*" Another gardener said: "*When the 'boss' comes home he don't wonder what I have been doing all day now. I like the Mower because it shows so much work done, and runs so easy.*" Another one states: "*We only used to keep the lawn cut around the house for a ways, and let the rest grow up in grass, but now I can mow the whole place in half a day, and I tell you it makes our grounds look fine.*"

A wide-cut Mower, in addition to doing the work a great deal quicker, lasts much longer than a narrow-cut Mower, for there is less wear; in other words, if it requires 100 times around with a 12-inch Mower to cut a lawn, it will only take fifty times around to cut it with a 24-inch Mower.

The "Henderson Ball-bearing" is strictly a high-class Mower; we have not endeavored to make it "cheap," but the best we know how—from the highest quality of iron and steel, which has enabled us to very materially reduce the weight. The castings are lathe-turned and bored, and the journals machine-trimmed, which insures trueness in fitting and smoothness in working, with no irregularities to wear off and leave rattle and play. The knives are of the finest steel and temper, being made especially for our Mowers by Loring Coes, long renowned as the maker of the finest, though most expensive, knives in America. Our knives are sharp and cut like razors, and will keep so if the under knife is kept tightened up to the revolving knives, for then the blades wear to the same bevel, and, in consequence, are self-sharpening. Simply bear in mind that a Lawn Mower, like a pair of shears, cuts by one blade rubbing over another. With shears, if the blades are apart, they bend the cloth over and do not cut; with a Lawn Mower, if the blades are apart, they bend the grass over without cutting it, or pull it, loosening the roots. The remedy is the same in both cases—simply tighten up and bring the cutting blades together. With most Mowers this is a difficult operation, and hard to understand, but in our new Ball-bearing Mower we have so simplified this adjustment that even those ignorant of mechanics will have no trouble in always keeping our Mower in fine cutting order, and the lawn in uninjured condition.

The Henderson Ball-bearing Mower has drive wheels 11 inches high, a large, open, non-clogging cylinder placed far enough back to prevent any "kicking up" of the Mower when it strikes heavy grass. The spiral of the knives is correctly proportioned to give a continuous cut, leaving the lawn as smooth as if shaved, entirely free from the corrugated surface left by other Mowers. The handles are braced and reinforced to prevent twisting and breaking when one side of the Mower is held off the ground, and besides the handle is equipped with improved "grips," which keeps the hands in a natural position, gives more power, and is less tiresome than a continued grasp, with wrists twisted, on a horizontal grip. We wish to reiterate that our "Henderson Ball-bearing" Mower is made without regard to cost—to be the best, most durable, and the easiest working Mower on the market, and while the prices are necessarily a little higher than those of Mowers simply made to sell, yet in the end our Mower will prove much cheaper, aside from annoying delays and repairs experienced with cheap Mowers.

PRICES OF THE HENDERSON "BALL-BEARING" LAWN MOWERS.

Strictly net; no charge for packing or cartage.

15-inch cut	$10.00	21-inch cut	$12.00
18-inch cut	11.00	24-inch cut	13.00

Note.—The 24-inch cut renders it unnecessary for us longer to make the small size horse Mowers. It will be found very useful on large lawns, even if you have a horse Mower, as there are often times when the ground is too moist and soft to work a horse over it without marring the surface.

The "Henderson" Horse Lawn Mower.

THE LATEST AND BEST.

Side draught keeps horse off uncut grass.

Knives can be raised or lowered instantly.

This is the best horse-power mower manufactured, simple in construction, very durable, nothing but the very best steel and iron is put in it. It is quickly adjusted to cut high or low, and the arrangement for throwing in and out of gear and for raising the knives when passing over stones, rough places or roads is operated from the seat. The side draught keeps the horse on the cut grass and prevents trampling down the surface of the lawn as smooth as velvet. We

that which is to be mown. The revolving knives are "high speeded," making a continuous cut and leaving the surface of the lawn as smooth as velvet. We make it only in one size, 36-inch cut, which we have found from our long experience with lawn affairs to be the best size for horse power. If you want a narrower cut, we advise you to get our 24-inch cut Henderson Ball-Bearing High Wheel Hand-Power Mower, offered on page 52. This latter machine will be found very useful, even if you have a horse machine, as often when the ground is moist and soft it would mar it to work a horse on it.

Price for the Henderson Horse Lawn Mower, 36-inch cut, complete with seat, shaft and side draught, **$65.00.** (*Strictly net; no charge for cartage or boxing.*)

IMPROVED {BEST QUALITY} HORSE BOOTS.

To Prevent Horses from Sinking in Damp or Soft Ground.
$9.00 per set of four.

UNIVERSAL LOW-PRICE LAWN MOWER.

For the benefit of those who want a low-price mower, and for those who have but little grass to cut, we, this season, are offering the best cheap mower made. While the prices are as low as "department store" prices, yet this mower will be found much better, both in quality of knives and quality of work. These mowers, of course, do not compare with our Henderson High Wheel Ball-Bearing Mower, either in perfection of work, durability, nor ease of operation; yet, for a low price mower they are superior to most others on the market.

Prices of Universal Low-Price Lawn Mowers:

(*Strictly net; no charge for cartage or packing.*)

10-inch cut..................$2.50	14-inch cut..................$3.00	18-inch cut..................$3.75
12- " " 2.75	16- " " 3.25	

HENDERSON'S EASY BORDER MOWER.

Especially designed to cut the grass on borders—the narrow strips of sod, sometimes only a few inches wide—between flower beds and walks, where a side-wheel mower would have one wheel down in the gutter, and, in consequence, the mower would cut into the sod. In our Border Mower a large light cylinder extends the full width of the mower and prevents and keeps the machine level with the surface and does perfect work. This machine is also useful for regular lawn work. It is very light-running, and will give thorough satisfaction.

Prices for Henderson's Easy Border Mowers:

(*Strictly net; no charge for boxing or cartage.*)

10-inch cut............................$6.00	
12- " " 7.00	
14- " " 8.00	
16- " " 9.00	
18- " " 10.00	
20- " " 11.00	

THE ENGLISH GOLF MOWER.

This is an export mower, made in the United States, especially for the British Golf Links, and is made according to their designs. A great number are in use over there, and the replies to our inquiries indicate that they give thorough satisfaction. Our criticism of the mower was that it ran a little heavy, but, as the manufacturer states, "this mower is made to do the finest kind of work, the revolving cutter has five blades and is speeded very high, making a close clip." The small front roller and back open cylinder, both extending clear across the machine, keep it perfectly level, leaving a smooth and even surface. The height of cut can be regulated to suit.

Prices of the English Golf Mowers (*strictly net; no charge for boxing or cartage*):

10-inch cut..................$7.00	14-inch cut..................$9.00	18-inch cut..................$11.00
12- " " 8.00	16- " " 10.00	20- " " 12.00
	24-inch cut, for two men............$15.00	

PATENT
APPLIED
FOR.

THE "HENDERSON" WATER BALLAST ROLLER.

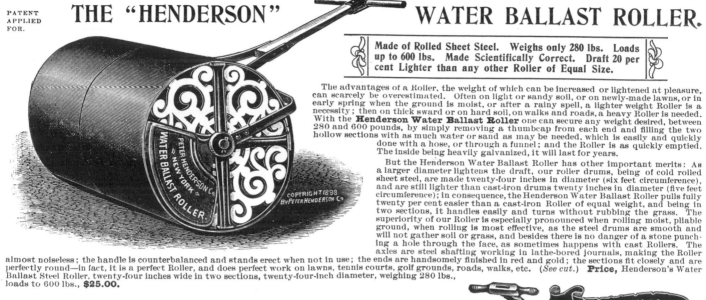

Made of Rolled Sheet Steel. Weighs only 280 lbs. Loads up to 600 lbs. Made Scientifically Correct. Draft 20 per cent Lighter than any other Roller of Equal Size.

The advantages of a Roller, the weight of which can be increased or lightened at pleasure, can scarcely be overestimated. Often on light or sandy soil, or on newly-made lawns, or in early spring when the ground is moist, or after a rainy spell, a lighter weight Roller is a necessity; then on thick sward or on hard soil, on walks and roads, a heavy Roller is needed. With the **Henderson Water Ballast Roller** one can secure any weight desired, between 280 and 600 pounds, by simply removing a thumbcap from each end and filling the two hollow sections with as much water or sand as may be needed, which is easily and quickly done with a hose, or through a funnel; and the Roller is as quickly emptied. The inside being heavily galvanized, it will last for years.

But the Henderson Water Ballast Roller has other important merits: As a larger diameter lightens the draft, our roller drums, being of cold rolled sheet steel, are made twenty-four inches in diameter (six feet circumference), and are still lighter than cast-iron drums twenty inches in diameter (five feet circumference); in consequence, the Henderson Water Ballast Roller pulls fully twenty per cent easier than a cast-iron Roller of equal weight, and being in two sections, it handles easily and turns without rubbing the grass. The superiority of our Roller is especially pronounced when rolling moist, pliable ground, when rolling is most effective, as the steel drums are smooth and will not gather soil or grass, and besides there is no danger of a stone punching a hole through the face, as sometimes happens with cast Rollers. The axles are steel shafting working in lathe-bored journals, making the Roller almost noiseless; the handle is counterbalanced and stands erect when not in use; the ends are handsomely finished in red and gold; the sections fit closely and are perfectly round—in fact, it is a perfect Roller, and does perfect work on lawns, tennis courts, golf grounds, roads, walks, etc. (See cut.) **Price,** Henderson's Water Ballast Steel Roller, twenty-four inches wide in two sections, twenty-four-inch diameter, weighing 280 lbs., loads to 600 lbs., **$25.00.**

HENDERSON'S LOW-PRICED CAST-IRON HAND ROLLER.

CAST-IRON ROLLER

For use on the lawn, a "Two Section" Roller is usually chosen, as it can be turned without injuring the grass. "One Section" Rollers will be preferable for walks, tennis courts, etc., as they leave no mark. About 300 lbs. is the weight chosen for one-man power for lawns. The handles of our Rollers are weighted so they always keep up from the ground, clean and out of the way. Our new Steel Lawn Rollers are the best, but if a cheaper Roller is wanted, our "Cast-Iron" will give good satisfaction. (See cut.)

		Diameter		length			in	sections,	weight			Net Price,
No. 1.		15 inches		15 inches		15 inches in	2 sections,		125 lbs.			$5.50
No. 2.		20 "		16 "		2 "			225 "			9.00
No. 3.		20 "		20 "		1 "			250 "			10.00
No. 4.		20 "		24 "		3 "			300 "			12.00
No. 5.		24 "		24 "		3 "			400 "			16.00
No. 6.		24 "		32 "		4 "			500 "			20.00
No. 7.		28 "		24 "		3 "			500 "			20.00
No. 8.		28 "		32 "		4 "			625 "			25.00

Horse Lawn Rollers and Field and Road Rollers. See page 14.

HENDERSON'S "MODEL" HAND LAWN SWEEPER.

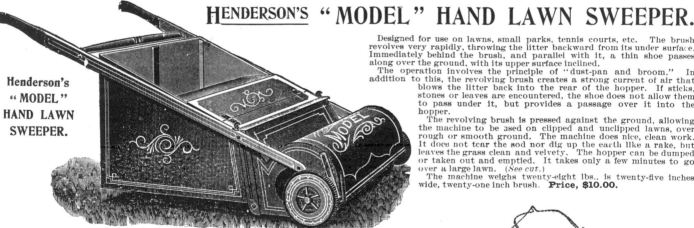

Henderson's "MODEL" HAND LAWN SWEEPER.

Designed for use on lawns, small parks, tennis courts, etc. The brush revolves very rapidly, throwing the litter backward from its under surface. Immediately behind the brush, and parallel with it, a thin shoe passes along over the ground, with its upper surface inclined.

The operation involves the principle of "dust-pan and broom." In addition to this, the revolving brush creates a strong current of air that blows the litter back into the rear of the hopper. If sticks, stones or leaves are encountered, the shoe does not allow them to pass under it, but provides a passage over it into the hopper.

The revolving brush is pressed against the ground, allowing the machine to be used on clipped and unclipped lawns, over rough or smooth ground. The machine does nice, clean work. It does not tear the sod nor dig up the earth like a rake, but leaves the grass clean and velvety. The hopper can be dumped or taken out and emptied. It takes only a few minutes to go over a large lawn. (See cut.)

The machine weighs twenty-eight lbs., is twenty-five inches wide, twenty-one inch brush. **Price, $10.00.**

PHILADELPHIA HORSE LAWN SWEEPER.

A revolving brush in front sweeps the sod clean and throws the sweepings backward into a large box in the rear. This box can be dumped without stopping. Sweeps forty inches wide.

The sweeper gives you a perfect lawn as soon as the sweeping is finished, instead of looking gray for several days owing to the withered cut grass.

If the sweeper and lawn mower are run in opposite directions, the lawn will be the same shade of color all over. **Price, $70.00.**

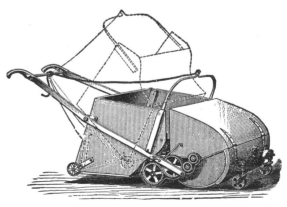

PHILADELPHIA HORSE LAWN SWEEPER.
DRAFT ROD NOT SHOWN.

EUREKA LAWN....
MOWER SHARPENER.

A very useful little implement for putting a sharp edge on the revolving cutter knives. It will not grind out nicks, nor file the machine, but will "whet up" the knives in short order, as you would a scythe or carver. You turn the machine backwards, holding the emery surface against the blades. It whets at the right bevel and all knives evenly. Price, with full directions for operating, $1.00 each.

REVOLVING "VERGE" CUTTER.

The "Planet, Jr." Lawn and... Turf Edger.

This little tool trims the turf around the edges of the flower beds, walks, roads, etc.; the revolving cutter does perfect work, either on a curved or straight border, edging accurately at just the correct angle, and at the speed of a mile an hour, while the hoe cleans the bottom of walk. The hoe can be removed if desired. It is invaluable in giving a finished appearance to the lawn. Weighs 26 pounds. **Price,** *complete,* **$5.00.**

Revolving Verge Cutter.

A splendid tool for trimming the turf around the edges of walks and flower beds. A revolving knife enables the work to be done with rapidity. (*See cut.*)

To Operate:—The forward hand, holding the tool, should rest against the forward leg just above the knee, and the knee should assist in pushing the trimmer (as is sometimes done in shoveling earth); this prevents the knife from cutting in too deep.

Price, - - - **$1.00.**

THE "PLANET, JR." LAWN AND TURF EDGER.

The Richmond Sod Cutter.

This new implement cuts the sod of uniform width and thickness, in any length, so that it is particularly adapted to giving solidity to slopes in cuts and on embankments, working equally well on level or uneven surface, cutting both ways with the land, and leaving no sod uncut, being light of draft, easily managed, strong, neat and durable.

One machine will cut from 30,000 to 40,000 square feet per day, thus doing the work of forty men.

We guarantee the expense saved between cutting by hand and with horse power in one day to pay the price of our machine.

Instructions for operating furnished with every machine. Every cutter warranted.

Prices:—8-in. Hand Machine, **$15.00**; 12-inch Horse Machine, **$30.00**; 14-inch Horse Machine, **$32.00.**

Is your....
Lawn Poor,
Worn Out
and full of
Bad Spots ?

Then it needs fertilizer—which can be broadcasted on evenly, and at any time of the year, with

The Stevens Hand Fertilizer Sower,

Which does the work perfectly, sowing all kinds of lawn dressing and other commercial fertilizers, wood ashes, lime, etc., in large or very small quantities. It can also be successfully used for fertilizing strawberry beds, and other garden and field work; is excellent for sifting wood or coal ashes, sawdust, or dry sand on icy walks; will also sow damp sand by removing two blades.

The hopper is 34 inches long and holds one and a half bushels. Weight of machine, 83 pounds. It is well made and nicely painted. **Price, $12.00.**

The Lawn and Garden Hand-Cart.

A most convenient cart. Is far superior to any wheelbarrow for all kinds of work.

Having two wheels, it is self-supporting when in motion, and the operator does not have to hold it from turning over sidewise.

It has large wheels, 32 inches in diameter, which are placed well under the box, so that the wheels carry nearly all of the load instead of the man carrying about one-half of it. Box is 3 feet 10 inches long, 21 inches wide inside bottom, and 15 inches deep.

The wheels being high, it can be handled in soft ground or mud and pushed or pulled over obstructions.

Price, - $8.00.

The Shuart Lawn Grader.

···· FOR SURFACING ····

Lawns, Parks, Athletic Grounds, and Tracks and Driveways.

This machine is especially adapted for fine grading, producing a beautiful even surface, and is thoroughly practical. It was originally devised for leveling for irrigation, and it does its work with such perfection that water will spread over acres by the mere influence of gravity.

It is a very efficient machine for conveying and spreading earth. The earth can be rapidly shaved down to the exact grade and contour desired. The construction of the machine is such that in the process of gathering, conveying and spreading the operator has complete control over the earth and can produce precisely the effects desired.

The weight of the machine is 440 lbs.; length of blade, 5½ feet; width of blade proper, 14 inches, to the top of which is fastened a supplementary blade 6 inches wide; has a capacity for three horses, but can be used with two horses without overloading. **Price, $48.00.**

Henderson's Large Lawn Barrow.

A very large barrow, designed to hold a large quantity of leaves, litter or manure. It is strong and will stand heavy work. Box, 25 inches wide by 32 inches long by 18 inches high. Nicely finished.

Price, - - $5.00.

(*For other styles Wheelbarrows refer to index.*)

···· HENDERSON'S ····

Lawn and Leaf Hand-Cart.

Especially adapted for carrying large quantities of light material, such as cut grass, leaves and litter; the box is deep and flaring—measures at the bottom 28 inches wide by 48 inches long, 10 inches wider at the top; it is 24 inches deep; nicely painted green and vermilion, striped and varnished; 30 inch wheels.

Price, - $16.00.

Lawn Sprinklers

A GOOD Lawn Sprinkler is a necessary adjunct to all well-kept lawns and grass plots. The water is distributed finely and evenly, and by leaving a sprinkler in one position long enough the grass can be thoroughly saturated, thereby insuring a luxuriant, fresh green sward during the hottest summer drought. All of the sprinklers that we catalogue work satisfactorily with a water pressure of 30 lbs., excepting the **Henderson 8-Arm Lawn Sprinkler,** which should have a pressure of not less than 40 lbs. The **"Water Witch"** and **"Niagara"** sprinklers work under a very low pressure better than any other sprinklers we know of. All sprinklers are fitted to attach to the regulation ¾-inch bore garden hose.

We are frequently asked if a tank or pond of water at a certain elevation will work a lawn sprinkler. The following rule is applicable to all such cases, viz.: Water at every foot of elevation gives half a pound pressure. The force to work a sprinkler requiring 20 lbs. pressure, the tank or body of water would have to be elevated 40 feet, etc.

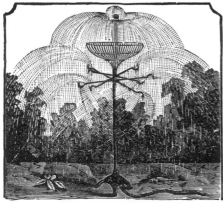

THE HENDERSON LAWN SPRINKLER.

COMMON SENSE LAWN SPRINKLER.

HENDERSON'S ("4-ARM" and "8-ARM") LAWN SPRINKLERS.

This is the best lawn sprinkler made. It can be attached to hose and placed anywhere on the lawn, where it serves the double purpose of a beautiful fountain and of thoroughly saturating the grass or garden. The water is distributed in fine drops over an area of 60 to 100 feet in circumference, according to the pressure on the water. An ornamental standard, about four feet high, is surmounted with long perforated arms, which revolve, so the beautiful sprays are constantly changing. **Prices:** for a 4-arm sprinkler, $2.75; for an 8-arm, $3.25.

BALL and BASKET. This fits on top of the Henderson sprinkler, and the stream of water keeps the silvered ball dancing on its summit up in the air. When the ball falls the basket catches it, and it rolls into the centre and is again raised by the water. **Price, $2.50 extra.**

The "TRAVELLING" ... LAWN SPRINKLER.

This, under an ordinary pressure of water, will roll itself slowly and steadily in any direction, either in a straight line or in a circle of any diameter desired. It can be set to travel anywhere from 15 to 500 feet per hour, as preferred. Can be changed from the lowest to the highest speed in a moment. It can be gauged to throw the water over a swath anywhere from 5 to 50 feet in width. It can be used as a stationary sprinkler by throwing it out of gear. With a pressure of water as low as 20 pounds it will drag 75 feet of hose, while with a pressure of 60 to 70 pounds it will carry 200 pounds extra weight and drag as many feet of hose. It works to perfection—in fact, doing the work more thoroughly and in less time than a person could do it, and needs less attention. The Travelling Sprinkler, even when stationary, is equal to all others. **Price, $14.00.**

THE "COMET" LAWN ... SPRINKLER.

Will Sprinkle an Area FOUR TIMES Greater than any other Sprinkler.

By means of the swiftly revolving arms and intermediate gears the upper part of the body is made to revolve slowly around, carrying the hose nozzle, from which a large stream of water is thrown far out beyond that thrown by the arms, thereby covering four times the space of any other stationary sprinkler. With an ordinary force of water it will thoroughly sprinkle a space of 80 feet in diameter. The hose nozzle and tips on ends of arms are adjustable, so that it can be adapted to as small a space as desired. Excepting the legs, it is made of solid brass and nicely nickeled. Most substantial and durable as well as the most attractive and useful sprinkler ever made. **Price, $5.00.**

The "Comet" Lawn Sprinkler.

COMMON SENSE LAWN ... SPRINKLER. ...

There are no moving parts to get out of order, no arms to get bent, stopped up, etc. The Sprinkler may be cleaned in a moment, should it become choked; by simply loosening the nut, the force of water will drive out any dirt that has accumulated.

In use the most beautiful water effect is produced, there being a large umbrella spray, very fine, as well as jets nearly upright; these jets not only add to the beauty of the fountain, but by falling upon the umbrella spray break it up more and dash the water over all parts of the large circle. The parts all being rigid there is little loss of pressure.

While the area covered is very large, yet the spray is so fine that comparatively little water is used; not only this, but the spray being very fine, the sunshine produces most beautiful rainbow effects.

The diameter of the circle will vary with the pressure; good to heavy pressure will wet a circle 20 to 40 feet in diameter. **Price, $1.25 each.**

The "NIAGARA" ... LAWN SPRINKLER.

An unequalled low pressure Sprinkler. Under a pressure as low as 5 pounds it will distribute a spray of uniform density over an area 15 feet in diameter, and with a water pressure of 25 pounds or over it will cover an area of 30 feet or over in diameter. It is very unique and simple—with no revolving parts to wear or get out of order—and by simply turning a knurled sleeve any desired density of spray can be obtained.

Price (*with spur to stick in the ground*), 35c. each, or by mail, 40c. each.

THE "WATER WITCH" LAWN ... SPRINKLER.
(With or Without Stand.)

An effective device for sprinkling lawns, gardens or flower beds. The water flows with unimpeded force, and is divided and deflected by the two lips of the swivel piece, which it causes to revolve rapidly, scattering the water in fine drops and evenly over a circular area of 25 to 40 feet diameter. It works more satisfactorily with a very low pressure of water than any sprinkler we know of.

Price, without stand (*i. e., with spur to stick in the ground*), 75c. each; by mail, 85c. each.

Price, with stand (*cannot be mailed*) (*see cut*), $1.25 each.

THE BONNET LAWN SPRINKLER.

The BONNET LAWN ... SPRINKLER.

Will Sprinkle either a Half or Full Circle by simply turning the thumb nut.

We consider this one of the most durable and satisfactory sprinklers ever placed on the market. It sprinkles either a full or a half circle. When set to sprinkle a half circle it can be placed close to a walk, and while wetting the lawn yet allows one to pass on a dry walk. The distributing channels of this Sprinkler are large, and thus there is no chance to clog. It will sprinkle a circle from 2 to 40 feet, according to water pressure. The operation of changing from a whole to a half circle is so simple that a child can work it. For full circle place the revolving wings in the centre; for half circle, place them on one side.

Price (*with spur to stick in the ground*), $1.50; or by mail, $1.65.

THE "WATER WITCH" LAWN SPRINKLER.

HENDERSON'S "Siamese" Lawn Sprinkler Attachment

"SIAMESE" ATTACHMENT.

By Using These Several Sprinklers can be operated at one time,

Providing the pressure and volume of water are sufficient. A three-quarter inch hose and thirty to forty pounds pressure will operate three sprinklers; a forty to fifty pound pressure, four sprinklers; with one inch hose and a good pressure, six sprinklers can be operated satisfactorily. This method of watering is valuable for thoroughly saturating large areas of lawn or garden, distributing the water more evenly and thoroughly than can possibly be done with a hose nozzle.

We found the "Water Witch" Lawn Sprinkler (*with spur to stick in the ground*) the best sprinkler for the purpose. One of these sprinklers and the "Siamese" was applied at the end of every twenty-five foot length of hose. PRICE, "Siamese" Attachment for ¾ inch hose, 75c., by mail, 85c.; for 1 inch hose, 85c., by mail, 95c.

For prices of Water Witch Sprinklers see opposite page.

Lawn Sprinkler Carriage.

Can be used with any sprinkler, having a spur to stick in the ground—the carriage enables the sprinkler to be moved without shutting off the water. PRICE, 75c. each.

Brass Hose Nozzle.

Brass with stop cock in the large end, spray rose and straight stem. PRICES, for ½ and ¾ in. hose, 75c.; for 1 in., 85c. Postage, 10c. each extra.

The "Graduating" Spray Nozzle.

Will throw a coarse or a fine spray, a large or a small solid stream. The spray can be closely contracted or made to cover a large area. These results are obtained by revolving it part way round. PRICE, ¾ in., 50c.; 1 in., 60c. Postage, 5c. each extra.

Hudson's Hose Menders.

PAT'D MAY 7TH 1889

Practical, simple, perfect. PRICE, per box of 6 tubes, 20 bands and 1 pair pliers, 75c.; or by mail $1.00. (*Give size of hose.*) TUBES, ½ in. 2c.; ¾ in. 3c.; 1 in. 4c. each. BANDS, 20c. per dozen. PLIERS, 30c.; by mail, 35c.

Hose "Shut off."

Can be inserted in hose at any point, thereby saves much time. For ¾ in. hose, 90c., by mail, $1.00.

Gem Nozzle Holder.

For watering lawns and flower beds. The hose is held firmly and can be adjusted to any elevation in an instant. 25c. each, by mail 30c.

The "Rain Maker" Hose Nozzle.

"RAIN MAKER" HOSE NOZZLE.

It throws a flat, fan-shaped sheet of water which breaks up into myriads of small drops falling like gentle rain and covering an arc-like area of from six to twelve feet in diameter according to pressure. The spray is so evenly scattered that the soil does not pack nor wash out from the roots of plants or grass. For watering greenhouses, spraying under foliage as well as above it, it is unequalled, and its use is sure death to red spider. For sprinkling lawns it is especially valuable. It can be simply thrown down or used in connection with our hose nozzle holder (*offered below*). It is made of brass and is practically indestructible. There is no clogging, no leaking on the operator. Please state which of the undermentioned you want.

No. A, has 3-16 in. flow, for a medium water pressure, fits ¾ in. hose .... **75c.** each.
No. B, has ¼ in. flow for a heavy water pressure, fits ¾ in. hose, **75c.** each.
No. C, has ⅛ in. flow, for killing red spider or roses, etc., fits ¾ in. hose .... **75c.** each.
No. D, has 5-16 in. flow, for heavy water pressure, fits 1 in. hose, **$1.00** each.
(*If by mail, add 10c. each extra for postage.*)

The "Cooper" Brass Hose Mender.

PAT'D. SEPT. 22. 96.

Made of thin brass tubing; will not rust or wear out; scarcely decreases the flow of water. Easily applied by any one: simply cut out your bad piece of hose and force the ends of the good hose over the mender until they meet in the centre. No other fastening is required; the barbs will hold it firm, and no matter what strain is put on the hose it will be as good as new at the point mended, and will not leak. PRICE, for hose with ½ in. bore, 7c. each, 75c. doz.; for hose with ¾ in. bore, 7c. each, 75c. doz.; for hose with 1 in. bore, 8c. each, 85c. doz. If wanted by mail, add 5c. per doz., at the single price postage free.

Hydrant Swivel Connection.

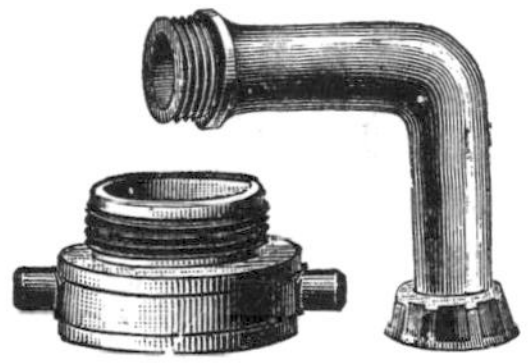

Turns in any position, prevents hose from kinking. For ¾ in. hose, 90c., by mail, $1.00. **Hose Reducer.** 1 in. to ¾ (*mailed free*), 30c.

Lightning Hose Coupler.

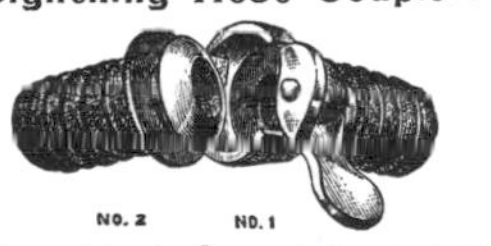
NO. 2
NO. 1

Instantly attached or disconnected, no twisting of the hose, no bruising fingers, water-tight. (*State if wanted to replace old couplings, or to be attached to them.*) PRICE, post-paid, per set two pieces, ¾ in., 30c.; faucet attachment extra, 15c.

Florists' and Gardeners' Hose Sprinkler.

A wide face nozzle. The holes are made small and are numerous, so that a copious yet gentle shower is given without washing or packing the soil.

PRICE, for ¾ in. hose, 3 in. face, 50c.; 4 in. face, 75c. Postage, 10c. each extra.

Caldwell's Hose Strap.

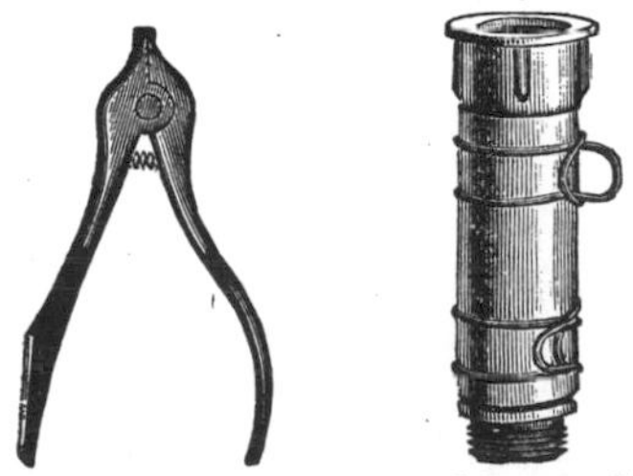

The best device for attaching hose couplings. Pliers, 30c., by mail, 45c. pair; hose straps for ½ in. hose, 50c. doz.; ¾ in. hose, 60c. doz.; 1 in. hose, 80c. doz.

HENDERSON'S WATER BARREL TRUCKS AND ACCESSORIES.

The **Barrel** is raised from the ground, carried to the place desired, and can be readily disconnected from or attached to the truck while barrel is either full or empty. We have wheels of 1½, 2½ and 3½ inch tire, *but always send truck with wheels 1½ inch tire, unless otherwise specified.* **We also supply as extra attachments the following: A Box** with trunnions and spring catch, making a very superior *dumping* Hand-cart. **A Sprinkler,** invaluable for watering lawns and sprinkling walks. Water is turned on and off by hand wheel and ball valve. **A Leaf Rack,** very useful for removing leaves and litter. (*Knocks down for shipment.*) **A Force Pump,** "The Gem," for washing windows and carriages, spraying trees, watering plants, etc., fully described on page 65.

COPYRIGHT 1897 BY PETER HENDERSON & CO.

PRICES.

Truck and barrel, 1½ inch tire......$10.00	Extra Trunnions, per pair......$0.50	Gem Force Pump Attachment.......$5.00
" " 2½ " 11.00	Hand-cart Box...... 2.75	
" " 3½ " 12.00	Leaf Rack...... 4.00	(*If truck and trunnions without barrel are wanted,*
Extra barrel, with trunnions on...... 2.50	Sprinkler Attachment...... 2.50	*deduct $2.00 from prices of trucks with barrel.*)

THE "LITTLE GEM"
HORSE POWER SPRINKLER.

"LITTLE GEM" HORSE POWER SPRINKLER.

For Sprinkling Lawns, Driveways, Trees, Gardens, etc.

Will spread water 18 feet wide, or the spread may be reduced to any desired width, down to 1 foot, at the will of the operator; or it can be readily adjusted to apply one or two narrow streams at one time, directly onto vegetables or other plants in rows, thus sprinkling two rows at one time. It is the only sprinkler adapted for spreading liquid manure, as it will not clog. The valves are operated from the driver's seat, thus giving him perfect control of the flow at all times, each side working independently of the other. Capacity, 150 gallons; tire, 4 inches wide; wheels, 4½ feet high; tracks (outside), 4 feet 10 inches wide; gear, painted red; tank, green; weight about 800 lbs.

Price, $90.00, net. (*List price, $125.00.*)

"JUNIOR" HAND POWER SPRINKLER.

"LITTLE GEM, JUNIOR"
HAND POWER SPRINKLER.

Operated by a Man or Boy as easily as a Wheelbarrow.

For sprinkling lawns, walks and drives in private grounds; also useful for laying the dust in large buildings and stables. Capacity, 52 gallons; weight, 280 lbs.

Price, $20.00, net. (*List price, $25.00.*)

"ALL IRON" HOSE REEL.

Is constructed entirely of iron, and is indestructible. It is light in weight, frictionless, and the wheels being high make it easily manipulated; a good, strong, handsome and convenient reel.

PRICES.

No. 10, 21 in. wheels, holds 100 ft. ¾ in. hose, $2.75
No. 20, 24 " " " 150 " " 3.00
No. 30, 30 " " " 500 " " 5.00

BENT LEG HOSE CARRIAGE.

This is the best wooden reel made—not easily tipped over. Can be rolled from place to place. By using a reel the hose is always drained, thus preventing it from rotting.

PRICES.

No. A, for 100 feet of ¾ inch Hose......$2.00
No. B, for 200 feet " " 2.25

HENDERSON'S "BEST PARA" HOSE.
(*Showing new "water-tight" couplings on.*)

HENDERSON'S "BEST PARA" RUBBER HOSE.

This is the highest grade of garden hose, being made entirely from New Para Rubber, and will outlast cheap hose three times over. We guarantee it to stand a 200-pound water pressure, and it will be as resilient in three or four years' time as when new, while cheap hose of that age will be hard and rotten. Every length of Henderson's "Best Para" Hose is fitted with the new water-tight couplings, without extra charge. (*Hose is furnished only in 25 feet and 50 feet lengths; the sizes ¾ and 1 inch refer to the internal diameter of the bore.*)

PRICES "BEST PARA."

¾ inch, per 25 foot length......$4.00; 50 foot length......$7.50
1 " " 25 " " ...... 5.50; 50 " " 10.00

"STANDARD QUALITY" RUBBER HOSE.

This is a grade that is often sold as the "best." While it is a good hose and will stand a 75-pound pressure, yet it is not warranted. Each length is fitted with the regular couplings, without extra charge.

PRICES "STANDARD QUALITY."

¾ inch, per 25 foot length......$3.00; 50 foot length......$5.75
1 " " 25 " " ...... 3.75; 50 " " 7.00

BENT LEG HOSE CARRIAGE.

"ALL IRON" HOSE REEL.

"Half-Moon" Grass and Turf Edging Knife.

For trimming the sod around borders of walks, etc. Price, with handle, 50c. each.

Combination Border Knife and Scuffle Hoe.

For edging up borders, for cutting and rolling turf, for cleaning walks; instantly changed from edger to scuffle hoe. Price, 7½ in. blade, 60c. each ; 9 in. blade, 75c. each.

Bass Broom with Scraper Hoe.

For lawns and walks. It can be used to push, pull, sweep, chip or dig.

12 inch	...4 rows of bristles			$0.75
14 "	...5 "	"		.90
16 "	...5 "	"		1.00
18 "	...5 "	"		1.15

English Turfing Spade.

For lifting sod evenly, rapidly and without breaking, $5.00.

Turf Cutting Axe, $2.25.

The "Knuckle Saver" Grass Hook.

Particularly adapted for trimming lawns, the **Raised Handle** enabling the operator's hand to clear the ground, at the same time giving a square cut. Made from **Solid Steel**, ground sharp.

Price, No. 2, 35c. ; No. 3, 40c. ; No. 4, 45c.

Imported English Riveted Back Grass Hooks.

Thin sharp blades, strengthened by a riveted back, light and rapidly handled.

No. 1, 50c. ; No. 2, 55c. ; No. 3, 60c. ; No. 4, 65c.

Combination Grass Hook and Border Trimmer.

A very handy tool for cutting grass and trimming up the border.

Price, 50c. each.

Serrated Grass Hook.

The peculiar wavy cutting edge makes it a very rapid and easy grass trimmer.

Price, 40c.

"Curved Handle" Grass or Sheep Shears.

Extra long, 7 inch blades. The handles are curved to prevent rubbing the knuckles on the ground.

Price, $1.00.

Grass Edging or Border Shears.

For trimming the overhanging grass around the edges of walks, etc. Price, 9 inch blades, $2.25, or with wheel, $2.75.

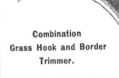

Lawn Shears.

For cutting grass under shrubs, fences, etc. Price, with two wheels, 9 inch blades, $2.50.

Hedge Shears, also used for Trimming Grass.

8 inch blades, $1.25, with notch,		$1.50
9 " " 1.50, " "		1.75
10 " " 1.75, " "		2.00

The "notch" is at the heel of the blades, for cutting small branches.

Wooden Lawn Rake.

Best selected three bow, 24 teeth, varnished head. 30c. each ; $3.25 per dozen.

Reversible Steel Lawn Rake.

Heavily tinned steel teeth. The arched teeth for leaves and litter, and the opposite side for fine cut grass. Twenty-four teeth closely set in a 21-inch head.

Price, 50c. each.

Wheel Lawn Rake.

The wheels enable it to ride over the lawn smoothly, without digging into the sod. The pointed teeth being adapted to grass and the back of rake reversed on handle, forming a perfect set of bowed teeth, principally adapted to raking leaves and rubbish ; 25½ inches long and has 28 teeth, tinned to prevent rusting.

Price, 75c.

Automatic Self-Cleaning Lawn Rake.

A backward motion of the operator will clean all the teeth at once, thus avoiding cleaning the teeth with the fingers. Made of best hickory.

26 teeth		$0.80 each
38 "		1.00 "
52 "		2.00 "

English Daisy Rake.

For running over lawns, tearing off disfiguring daisy and dandelion flowers, which are carried in the deep hooded head.

16 inch head			$3.00
20 " "			3.50
24 " "			4.00

Imported English Riveted-back Lawn Scythe.

Light and thin broad blades, strengthened by a riveted back, 32 inches, $1.20 ; 34 inches, $1.30 ; 36 inches, $1.40 ; 38 inches, $1.50.

Scythe Snaths or Handles.

With patent fastening. Price, 75c. each.

Henderson's Lifting Weeder.

A chisel blade cuts off the weeds ; the trigger holds them so they can be pulled out. 60c. each.

Henderson's Cane Weeding Gouge.

A very useful and convenient tool for cutting weeds out of lawns without marring the surface. Price, 50c. each.

CHISEL BLADE

Chisel Blade Weed Cutter.

With foot rest so it can be pressed, cutting off roots of weeds without marring the lawn. Price, 50c. each.

Henderson's Giant Lawn Umbrella.

Eight feet across.   Alternate stripes of bright colors.   Very showy on a green lawn.

GIANT LAWN UMBRELLA.

These giant framed umbrellas are the strongest and most durable umbrellas in the world. They are not only very useful but very attractive on a lawn; they spread 8 feet. The centre pole is thrust into a large hollow iron screw, which is easily turned into the ground; the screw can remain, if desired, and the umbrella can be closed and removed in a moment. It can be quickly set up in any soil with ease. The umbrella has 16 strong steel ribs covered with heavy cloth, in alternate bright colors, usually red, yellow, blue and buff; the edge is finished with an 8-inch curtain. Price, complete, with ground fastenings, $7.50.

LAWN SETTEES.

With or without Canopies.

..THE..
"COMFORTABLE" SETTEE.

This is the most comfortable seat imaginable; made of narrow strips of selected hard wood, varnished; heavy cast-iron legs.

No. 1, 5 ft. long, $9.00; with awning $19.00
" 2, 6 "   10.00; "   " 21.00
" 4, 7 "   12.00; "   " 24.00

Old=Fashioned Sun Dials.

Very unique and interesting lawn ornaments.
Price...............$2.50 each.

CEDAR CHAIR.

PARK SETTEE.

A very popular settee, with varnished natural wood seat and back, with painted iron work.

No. 1, 3½ ft. long..................................$3.50
" 2, 4½ " 4.00
" 3, 5½ " 4.50

"CHEAP" SETTEE.

An all-wooden settee, large enough for two persons; neatly finished.
Price..$1.75 each.

RUSTIC CHAIRS AND SETTEES.

Cedar Chair (*see cut*)...........................$8.00
Cedar Settee, 5 ft., $15.00; 6 ft...........................18.00

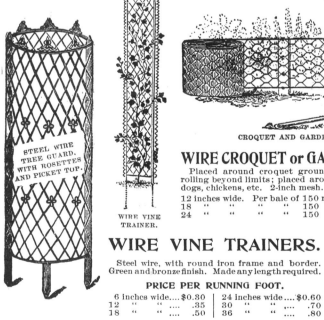

STEEL WIRE TREE GUARD. WITH ROSETTES AND PICKET TOP.

WIRE VINE TRAINER.

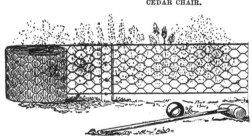

CROQUET AND GARDEN BORDER.

WIRE CROQUET or GARDEN BORDER.

Placed around croquet grounds prevents balls from rolling beyond limits; placed around flower beds excludes dogs, chickens, etc. 2-inch mesh. Galvanized.

| 12 inches wide. | Per bale of 150 running feet...........$2.50 |
| 18 " " " 150 " " .... 3.75 |
| 24 " " " 150 " " .... 5.00 |

WIRE VINE TRAINERS.

Steel wire, with round iron frame and border.
Green and bronze finish. Made any length required.

PRICE PER RUNNING FOOT.

6 inches wide.... $0.30 | 24 inches wide.... $0.60
12 "   " .... .35 | 30 "   " .... .70
18 "   " .... .50 | 36 "   " .... .80

WIRE FAN TRAINER.

WIRE LAWN ARCHES FOR CLIMBERS

WIRE FAN TRAINERS.

Steel wire. Green and bronze finish.

No. 9.	12 inches high...	$0.25 each,	$2.50 doz.
" 10.	15 " " ...	.30 "	3.00 "
" 11.	18 " " ...	.40 "	4.00 "
" 12.	24 " " ...	.50 "	5.00 "
" 13.	30 " " ...	.65 "	6.50 "
" 14.	36 " " ...	.80 "	8.00 "
" 15.	42 " " ...	1.00 "	10.00 "
" 16.	48 " " ...	1.50 "	15.00 "

TREE GUARDS.

Steel wire, 2-inch mesh. Finished in green and bronze. Made in halves, in any diameter and height, so they can be readily joined around the tree.

PRICES.

Plain, per square foot..................................$0.50
or with rosettes (*as shown in cut*)........... .60
Picket top extra (*as shown in cut*), per
lineal foot... .20

WIRE HANGING BASKETS.

Green and bronze finish; strong and well made. Just the thing for decorating piazzas of summer hotels and suburban residences.

No. 4, 8 inches in diameter...........25c. each, $2.50 per doz.
" 5, 10 "   "   "   ....30c. "   3.00 "

WIRE ARCHES.

For Lawns and Gardens.

Steel wire, with round iron frame and border. Green and bronze finish. Made any size required.

PRICE PER RUNNING FOOT.

6 inches wide .... $0.30 | 18 inches wide.... $0.50
12 "   " .... .35 | 24 "   " .... .60
15 "   " .... .40 |

CEDAR TREE TUB.

CEDAR TREE TUBS.

Heavy iron hoops. Drop handles, serving as hooks for carrying poles. Iron legs. Removable perforated bottoms. Painted green outside and brown inside. Painted red, if desired, at an extra cost of 10 per cent.

No.	Outside Diam.	Length of Stave.	Price.
0.	27 inches.	24 inches.	$6.00
1.	25 "	22 "	5.00
2.	23 "	20 "	4.50
3.	21 "	18 "	4.00
5.	18 "	16 "	3.50
6.	16 "	14 "	2.75
7.	14 "	12 "	2.25
8.	13 "	11 "	2.00
9.	12 "	10 "	1.75

"COLUMBIA" FLOWER TUBS.

Durable cypress staves, held by strong steel wire hoops, that can be tightened by draw screws connected with the handles. The "Fancy" (see cut) has staves finished like red wood, alternating with staves of natural finish, all varnished. The "Plain" has smooth top and is entirely natural wood.

No.	Diameter.	Height.	Fancy.	Plain.
A.	12 inches.	11 inches.	Each, $1.50	$1.00
B.	15 "	14 "	" 2.25	1.50
C.	18 "	16 "	" 3.00	2.00
D.	21 "	18 "	" 3.75	2.50
E.	24 "	20 "	" 4.50	3.00

FANCY "COLUMBIA" TUB.

Plant Stakes.

NICELY FINISHED.
PAINTED GREEN.

5 ft. long × 5/8 in. dia.
4 ft. × 9/16 in. "
3 1/2 ft. × 9/16 in. "
3 ft. × 1/2 in. "
2 1/2 ft. × 7/16 in. "
2 ft. × 3/8 in. "
1 1/2 ft. × 5/16 in.

2 ft. 1/2 in. Diam.
3 ft. 5/8 in. Diam.
4 ft. 3/4 in. Diam.
5 ft. 7/8 in. Diam.
6 ft. 1 in. Diam.

Dahlia Poles.

PAINTED GREEN.
WHITE TOPS.

STAKES, ROUND GREEN TAPERING.

	Each.	Per doz.	Per 100.
2 feet	3	$0.25	$1.75
3 "	5	.50	3.50
4 "	7	.75	5.00
5 "	8	.90	6.50

STAKES, SQUARE GREEN TAPERING.

	Each.	Per doz.	Per 100.
2 feet	3	$0.25	$1.50
3 "	4	.40	2.75
4 "	5	.50	3.50
5 "	6	.65	4.25
6 "	7	.80	5.50

DAHLIA STAKES OR POLES. ROUND, FANCY TURNED TOPS.

	Each.	Per doz.	Per 100.
2 feet	3	$0.30	$2.00
3 "	5	.50	3.50
4 "	7	.75	5.00
5 "	10	1.00	6.50
6 "	14	1.50	10.00

The Reddick Mole Trap.

The Daffodil MOLE TRAP.

THE DAFFODIL MOLE TRAP.

Especially designed for cold frames in which bulbs, roots, etc., are wintering, and other positions where a tall trap cannot be set. The Daffodil Mole Trap is simple and serviceable, made of steel. The needle bar is held, and locks automatically when setting the trap. The trigger is sprung by the mole passing under, and the needles are forced into the ground, killing the mole. Price, $1.00 each; 3 for $2.75.

THE REDDICK MOLE TRAP.

The best and most complete Mole Trap ever invented. An arrangement for holding the handle up renders it easy to set, and cannot "startle" or injure the operator while being set. It will catch moles when quite deep in the ground, and there being no pin or other obstruction projecting into the run, there is nothing to frighten or disturb the mole as it passes, and in doing so raises the ground over the "run" just enough to spring the trigger, which must firmly rest *on* the soil over the "run." The points of the pins being constantly in the ground, it cannot catch or injure chickens or other domestic animals. Cannot be blown over or injured in any way by rain or storm, and being made entirely of metal, cannot warp, twist or get out of order, and is light, neat and durable. The ground not being disturbed in any manner, it can be set close to plants without injuring them, and it can also be set touching a wall, fence, etc., without impairing the working of the trap in any way. Price, $1.25 each; 3 for $3.50; $13.00 doz.

WHERE TO SET A MOLE TRAP.

There are many tracks through which a mole passes but once, and, of course, it is useless to place the trap over such a run. To find a run which is frequently used, depress the ridges for a short distance in several parts of the lawn. As moles pass through some "runs" at regular intervals, say about 12 M., and again at 6 P.M., it is an easy matter to tell which "runs" are used by examining the depressions and noting those that have been raised. Full directions for setting sent with each trap.

RUSTIC HANGING BASKET.

These are now extensively used for decorating the plazzas of summer hotels and country residences.

PRICES.

8-inch	$1.00
10- "	1.25
12- "	1.50
13- "	1.75

"THE HENDERSON (ODORLESS QUICK-ACTING) LAWN ENRICHER."

IT NEVER FAILS TO INDUCE A RICH GREEN AND LUXURIANT GROWTH IN A FEW WEEKS' TIME.
CAN BE APPLIED IN SPRING, SUMMER OR FALL.

A clean, portable and convenient lawn dressing which is never-failing in inducing a rapid and rich green growth. It should be sown broadcast in the spring or fall, though it can be put on during the summer without injury or danger of burning the grass, and a remarkable improvement will soon be observed. It is in every way more desirable than manure, which is so often full of weed seeds. The use of "Henderson Lawn Enricher" entirely does away with the old practice of top-dressing lawn with stable manure, which was so objectionable on account of unsightliness and disagreeable odors.

Quantity Required. A 10-lb. package is sufficient to go over an area of 300 square feet, or for forming a new lawn, from 1,000 to 1,500 lbs. per acre, or a smaller quantity for renovating an old one. PRICE, 5-lb. pack., 25c.; 10-lb. pack., 45c.; 25-lb. bag, $1.00; 50-lb. bag, $1.75; 100-lb. bag, $3.25; per ton of 2,000 lbs., $50.

LYON'S AUTOMATIC LAWN FEEDER.—(*Odorless.*)

It applies the fertilizer while you Water, and Feeds your Grass or Garden with High-Grade, immediately available, Concentrated Plant Food....

It is a neat, light device, attached between the end of the hose and the nozzle or lawn sprinkler and automatically dissolves a stick of *Concentrated Fertilizer*, diffusing it through as much water as will pass through an ordinary garden hose in about one hour. It is clean, odorless, non-poisonous and will give perceptible results after one week's use on your lawn or in your garden. The Cartridges that fit into the feeder are prepared from the *most powerful and efficient concentrated fertilizers.* The analysis shows 4½ per cent of Phosphoric Acid; 13½ per cent Potash; 12¼ per cent Ammonia, a combination to make the plant life grow and flourish, keep the grass green, and the flowers in bloom. The fertilizer is sufficiently hardened to prevent a too rapid wasting, but is thoroughly dissolved as it leaves the nozzle and is therefore in a condition to be readily absorbed by the roots of plants or grass. Complete outfit, consisting of one holder and 12 of the food cartridges, prepaid to any address on receipt of $1.00.

Extra Food Cartridges, in boxes of 24 for 50c., or prepaid for 75c.

Make your Garden and Grass Grow by using on your hose

STOTT'S FERTILIZER DISTRIBUTER.

Illustrated below.

Fill the cells with the Henderson Garden Fertilizer or Lawn Enricher (*offered on page 69*), or any good commercial fertilizer, which will be gradually dissolved and diffused by the water passing through the Distributer. Fertilizer applied in this manner is soluable, comes quickly in contact with the roots and is immediately available as a plant food; the effects are quick, luxuriant and healthy growth. It is valuable for use in flower gardens, vegetable gardens, for lawns and grass plots, putting greens, etc., where top dressing is frequently neglected, this distributer will be found especially beneficial. It can also be used for applying such insecticides as Whale Oil Soap, Tobacco Soap and Fir-tree Oil Soap (*offered on page 62*).

The Stott Fertilizer Distributer is an oblong cylinder machine made of copper, divided by perforated divisions into cells, into which the fertilizer or insecticide is inserted. At each end of the machine hose is attached, one end connecting with the water tap or garden pump, and the other to an ordinary hose nozzle or lawn sprinkler, the water being forced through, and is consequently impregnated with the fertilizer. Being continuous in action much time is saved, and on putting the composition in the celled divisions, it is immediately ready for use.

PRICES.

2 Cell,	- - -	$3.00
3 "	- - -	4.00
4 "	- - -	5.00

THE STOTT FERTILIZER DISTRIBUTER.

LIQUID MANURE MIXER AND ATOMIZER.

In a barrel place a liberal quantity of well-rotted manure, then fill it two-thirds full of water—if it can stand for a day or two, all the better. Connect the atomizer between two sections of hose and lower it into the barrel of manure water; close the nozzle so the water will pass down the suction or into the barrel. This will exhaust the air. Then open the nozzle, and the fertilizer will be mixed with water. To get large proportion of fertilizer, use check with large opening; to get small proportion, use check with small opening. It works admirably; mixes and distributes it thoroughly, evenly, rapidly and easily. This atomizer is also useful for the florist to mix or temper hot and cold water under water pressure.

Price, - - - $1.75 each.

Liquid Manure .... Mixer and Atomizer.

HENDERSON'S FLUID WEED DESTROYER.

The Best, Cheapest and Safest Destroyer of Weeds, Moss, Grass, etc., in Garden Walks and Carriage Drives....

It completely destroys all weeds, wherever applied, and by its effects on the ground prevents the growth of fresh ones for a year afterwards, thus saving an immense amount of labor in hoeing, etc.

Its application is easy, being in a liquid form and only requiring to be mixed with water and applied with a watering can. Four gallons mixed with a hundred gallons of water will cover an area of about 50 yards square. Full directions for use with each package.

Quart Can, *sufficient to make 6 gallons of liquid,* **50c.**
Gallon Can, *sufficient to make 25 gallons of liquid,* **$1.25.**
5-Gallon Keg, *sufficient to make 125 gallons liquid,* **$4.00.**

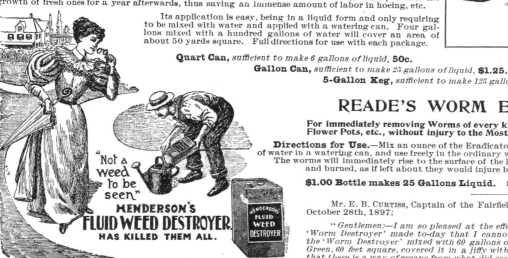

"Not a weed to be seen"

HENDERSON'S FLUID WEED DESTROYER.
HAS KILLED THEM ALL.

READE'S WORM ERADICATOR.

For immediately removing Worms of every kind from Lawns, Garden Plants, Flower Pots, etc., without injury to the Most Delicate Plant or Flower....

Directions for Use.—Mix an ounce of the Eradicator (about two tablespoonsful) with one gallon of water in a watering can, and use freely in the ordinary way of watering.
The worms will immediately rise to the surface of the lawn or pot, and should at once be picked up and burned, as if left about they would injure birds or fowls.

$1.00 Bottle makes 25 Gallons Liquid. $3.00 Bottle makes 80 Gallons Liquid.

Mr. E. B. Curtiss, Captain of the Fairfield County Golf Club, Greenwich, Conn., writes October 28th, 1897:

"*Gentlemen:—I am so pleased at the effect produced by the first application of your 'Worm Destroyer' made to-day that I cannot refrain from writing you. Three quarts of the 'Worm Destroyer' mixed with 60 gallons of water sprinkled on the surface of a Putting Green, 60 feet square, covered it in a jiffy with countless squirming nuisances. We feel now that there is a way of escape from what did seem an incurable evil.*"

Egg Tester.
Fits on any lamp using an ordinary No. 1 burner or No. 2 sun burner, 35c.; or with lamp, 85c.

Unrivaled Incubator Thermometer.
Sure to remain on the egg, and gives the accurate temperature. Price, $1.00.

Incubator Thermometer.
With stand...................$1.00
Without stand................. .50

Moisture Gauge.
For testing the amount of moisture in incubators. $2.00.

"Medicol" Nest Egg.
A perfect imitation of a natural egg, yet is a perfect disinfectant and will drive away lice and other vermin.
Price, 10c. each, $1.00 per doz.

Pineland Drinking Fountain.
Galvanized iron. Light and self-feeding. Chicks cannot be drowned nor water polluted.
Large, for fowls and chicks, $0.75
Small, for chicks............ .35
Special, for pigeons......... .75

"Sanitary" Drinking Fountain for Poultry.
The best made. Cleaned and filled in a minute.
1-gallon, 50c.; 2-gallon, 85c.

Hallock's Food or Water Holder.
Of galvanized iron. Will not rust. Cannot be turned over. Contents cannot be polluted. Top quickly detached for cleaning and filling. 2-gal. size, 75c.; 3-gal., $1.00.

Swinging Feed Tray for Poultry.
NO WASTE, NO DIRT.
Hens cannot scratch the food out or the dirt in. Weight of hen tips the frame and throws her off. Upper rod tips up also if hen tries to roost over pan. Tray always right side up. Legs fasten anywhere.
18-in., $1.00; 27-in., $1.25; 36-in., $1.50.

Non-freezing Poultry Fountain.
Absolutely safe. Cannot be overturned. The water is kept warmed by a fireproof lamp.
Price, complete, $2.50.

Poultry Bit.
Will stop feather pulling instantly. Fowls can eat and drink with them on. Made to fit all breeds. 10c. each, $1.00 per doz.

Roup Syringe.
Fig. 1 shows its use internally. Fig. 2, externally. One of the best cures for roup and similar diseases to which poultry are so liable.
Postpaid, with instructions, 15c.

... FOR ...
Clover Cutters, - 27
Meat Choppers, - 43
Bone Cutters, - 27
Poultry Fencing, 36
Corn Shellers, - 28

Gape Worm Extractor.
Price $0.25

Progressive Poultry-killing Knife.
50c. each.

French Poultry Killing Knife. 50c.

Poultry Marker.
Marks two sizes, No. 1 and No. 2. 25c. each.

Cast-Iron Mortar.
For crushing shells, crockery, glass, etc. Price, $3.50 each.

Costellow Grit Machine.
Will crush granite, gravel, crockery, shells, etc., fine or coarse, simply by working the handle high or low. Price, $3.50.

The Trowbridge Clover Cutter.
The best small machine for cutting clover (green or dry) for poultry. Cuts back and forth. Two cuts to a complete motion. Cuts enough for fifty fowls in one minute. Can be gauged to cut from ¼ inch up. Machine weighs 7 lbs. Price, $3.00.

Pilling's Caponizing Set.
Double your profits by caponizing chicks. The operation is very simple—the instructions so full and explicit that any man, woman or child will be able to perform the operation.
Price of Caponizing Set.
In velvet-lined case, with "Complete Guide".....................$2.75

Aluminum Leg Bands.
These bands can be shortened to fit any bird. They do not cut or chafe. Easily applied and removed. 30c. per doz., $2.00 per 100; numbered or lettered.

Wire Hens' Nests.
Steel wire, japanned. They afford no place for vermin, and allow the air to circulate freely through the nest. Nest 14 inches in diameter, mesh 1½ inches; 20c. each, $2.00 a dozen.

Anti-Vermin Perch Brackets.
The cup and bracket are of iron, made in one piece. Keep the cups filled with kerosene, sulphur or insect powder, which prevents insects from the walls, etc., getting at the chickens at night.
Price, 50c. per pair.

Lice-killing Machine.
Will kill lice on chicks, pigeons, old fowls or turkeys, without injury to the birds.
Chick size $2.50
Standard size. 3.00
Turkey size............ 4.00

The Hoff Patent Poultry Coop.
A perfect hen coop. Little chicks can be fed without interference from the hen. All are secure from invasion of enemies. Food box detachable. Water cannot be upset. Bottom removable. Roof raises. Dimensions, 24x30 inches; 18 inches high in front. $3.50.

Safety Egg Carrier.
Just enough spring to prevent the least breakage. Open on the ends, preventing eggs from becoming musty. PRICES.
(Include Trays and Case combined.)
30 doz. size, $2.50 | 9 doz. size, $1.50
16 " " 2.00 | 6 " " 1.25
12 " " 1.75

THE NEW PINELAND INCUBATORS. IMPROVED AND UP TO DATE.

"PINELAND PEARL."

THE....
"PINELAND PEARL"

is a little machine especially adapted to hatching a few eggs at a time. It can be placed in the room of a house, if desired, and will be found a valuable little machine for amateurs or those who wish to raise a few of some special breed of fowls or birds.

PINELAND 100-EGG INCUBATOR.

These we consider the very best incubators on the market, being surpassed in no one particular by any other make, while they are far superior in all respects to most others. They are thoroughly reliable, with improvements strictly up to date, the outcome of years of experience by one of the oldest and most reliable manufacturers in the country, who guarantee every claim made for their incubators and brooders. There are no experimental theories worked into these machines. Every feature is true and tried.

Construction. Thoroughly seasoned No. 1 lumber, no shrinking, no warping. Exterior case, 1 inch Carolina pine, natural finish, varnished; interior case, 1 inch white pine, with an inch space between them packed with non-conducting air-tight material, thus making the walls 3 inches thick. The doors are of double glass, and great outside variations do not affect the temperature in the incubator.

Improved Deep Egg Chamber allows ample space in front and below the egg tray for young hatchlings to drop down and remain in the nursery compartment for a few hours, without opening the door at the critical period, while the balance of the eggs are hatching out. This prevents the first chicks from running over and interfering with those just getting out of the shells.

THE NEW PINELAND INCUBATORS.

Ventilation and Moisture. The outside air, which should be fresh and pure, enters the machine from the bottom through two tubes, one each side of the lamp flue, where it is warmed; from there it passes over the water pans, absorbing moisture, it then circulates through the chamber around and through the eggs, and passes out of the ventilators at the opposite end. When the water pans are raised into more heat, the air absorbs more moisture, if lowered into cooler air, they give off less moisture, or if water is omitted or pans removed the eggs can be dried down. Thus the three essential factors in hatching eggs artificially—even warmth, fresh air and proper moisture—are all under perfect automatic control, adapting this machine for all climates and all conditions, and all changes are done outside of the egg or hatching chamber, so there is no disturbance of temperature by opening the doors at critical periods.

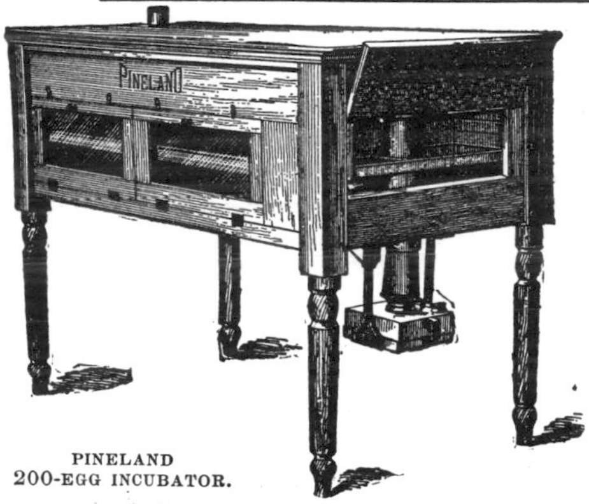

Equipment of New Pineland Incubators.....

Turning trays are supplied so a full tray full of eggs can be turned instantly without handling each egg individually. Non-breakable metal lamp and extra wicks, egg tester and thermometer are supplied with each incubator; also full directions in detail for operating.

C. L. DARLINGTON, POULTRYMAN, at HON. J. J. ASTOR'S, FERNCLIFFE POULTRY FARM, RHINEBECK, N. Y., reports Jan. 6th, 1897: "We have been using one of your Pineland 320-egg machines, and it gives entire satisfaction, having hatched 300 strong, healthy chicks out of 320 fertile eggs. Can we ask for anything better?"

Our Poultry Supply Department

is in charge of a gentleman who is a thoroughly practical, experienced and successful raiser of chickens and other poultry by artificial methods, and any points or information on this subject will be cheerfully imparted free to our customers; and if there are any poultry supplies wanted that we do not catalogue herein, we will, if possible, procure and supply them at lowest market prices.—P. H. & Co.

PINELAND 200-EGG INCUBATOR.

Heat : Improved Application and Regulation.

The heat from the automatic self-regulating lamp at the end of the machine circulates uniformly from the radiator throughout the chamber, passing out of a flue at the other end. The thermostat, which regulates the lamp and consequently the heat, is made of a composition that insures perfect mechanical regulation; the metal thermostats used in other machines, while they expand with heat satisfactorily, yet contract very indifferently, often causing serious results. These machines, properly set, cannot be overheated, and the lamps are perfectly safe.

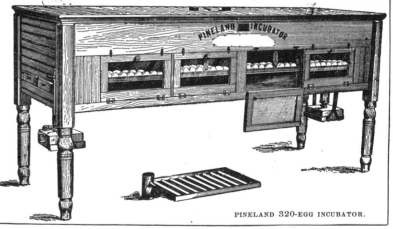

PINELAND 320-EGG INCUBATOR.

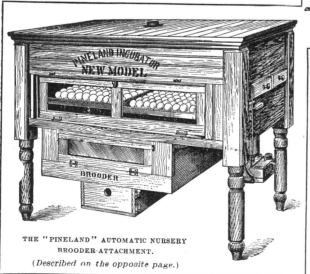

THE "PINELAND" AUTOMATIC NURSERY BROODER-ATTACHMENT.

(*Described on the opposite page.*)

PRICES OF NEW PINELAND INCUBATORS.

Full directions for operating are sent with each Incubator.

CAPACITY.	LENGTH.	WIDTH.	HEIGHT.	CRATED WEIGHT.	P. H. & CO.'S NET PRICE.	MAKER'S PRICE.
30 Eggs.	21 inches.	14 inches.	15 inches.	35 lbs.	$9.50	$10.00
50 "	30 "	18 "	32 "	75 "	15.00	18.75
100 "	3 ft.	2¼ ft.	3⅓ ft.	155 "	20.00	25.00
200 "	4⅙ "	2½ "	3⅓ "	225 "	32.00	40.00
300 "	6 "	2½ "	3⅓ "	250 "	44.00	55.00
*320 "	6¾ "	2½ "	3⅓ "	300 "	48.00	60.00
*400 "	8⅓ "	2½ "	3⅓ "	400 "	56.00	70.00

* Note. The 320 and 400 egg machines are practically two incubators in one, having two chambers and two lamps, one at each end of the machine. The separated chambers being operated independently, may be put into operation at different periods, or be used one for chickens and the other for ducks, or other poultry, as desired.

New Pineland Brooders. Improved and up to date.

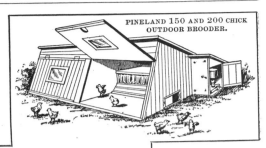

PINELAND 150 AND 200 CHICK OUTDOOR BROODER.

All Brooders are supplied with a non-breakable metal lamp, with improved burner. No chimney required.

The New Pineland OUTDOOR Brooder.
The best brooder on the market for outdoor service, adapted to both fanciers and practical poultrymen who raise a limited number of chickens. It is supplied with the most efficient brooding system known; is built of one-inch lumber; has no narrow inclined boards for chickens to enter a small aperture or door, but sits firmly on the ground. It has a side entrance, which prevents all draught from striking the young chickens in the brooding room, and no oil, smoke, chimneys or boilers pollute or encumber this room. No space underneath for chicks to huddle and freeze to death. It is snug and proof against wind and rain. There is no slanting roof to the opening of the brooder to make the ground wet and muddy for the young chicks. It was conceived by a practical poultryman, improved and perfected by others equally practical, and stands to-day without a peer as a successful brooder of chickens. It is supplied with a lamp having an improved burner, requiring no chimney. The 100-chick size has a brooding space of 672 square inches under the heater—the largest brooding capacity of any brooder on the market.

NEW PINELAND 50 AND 100 CHICK OUTDOOR BROODER.

The New Pineland INDOOR Brooder.
This brooder is divided by a partition, making it a brooder for two flocks of 50 to 75 each side. The partition acts as a ventilator, as it is made of two strips running lengthwise, about one inch apart. The fresh air enters this space at the lamp end and passes into each brooding chamber through holes bored in the partition. The roosting floors can be adjusted to the size of the chicks—at first near the tank and lowered as the chicks grow and need more room and less artificial heat. The heat is uniform in all parts of the brood chamber, being warmed by circulation of water through a tank the full width of the brood chamber, which obviates crowding or danger of smothering.

Our Pineland "Junior" Indoor Brooder
will mother 100 small chicks, and is almost the counterpart of the above, with the exceptions that it has but one large brooding chamber, devoid of the partition of the Indoor Brooder, and is somewhat smaller.

Our Poultry Supply Department

Is in charge of a gentleman who is a thoroughly practical, experienced and successful raiser of chickens and other poultry by artificial methods, and any points or information on this subject will be cheerfully imparted free to our customers. If there are any other poultry supplies wanted not herein catalogued, we will, if possible, procure and supply them at the lowest market rates.—P. H. & Co.

PINELAND 150-CHICK INDOOR BROODER

The Pineland Automatic Nursery Brooder.

(See cut on opposite page.)

The Pineland Automatic Nursery Brooder is attached to the bottom and directly underneath the incubator, and its operations may be thus described: As soon as the chickens are hatched they crowd to the front of the incubator where the light enters the glass door. The front end of the egg-tray is open, and the chicks tumble to the bottom of the incubator, where they alight upon a tilting board, so balanced that the weight of the chick causes it to sink or tilt on one side, and the astonished "peep" slides safely down an inclined plane into the brooding chamber below. Here he lands gently upon a layer of some soft material, and, in a strong, vigorous condition, is in the world ready to take care of himself. Remember, it is better never to open an incubator during a hatch; with the Pineland Automatic Nursery Brooder it is never necessary. This is the only strong point which really makes a nursery brooder valuable. No other hatcher or brooder has it. Our chicks don't have to be educated to climb upstairs; they go down-stairs by gravity. The Nursery Brooder is heated by a lamp, as in all our brooders, and when ready to use it the only thing necessary is to start up the heat according to directions.

Prices of Automatic Nursery Brooders:

For 100 and 200 egg incubators, $10.00 each.

For 300, 320 and 400 egg incubators, $12.00 each.

The New Pineland Indoor Sectional Brooders.

These are the latest and best inventions for raising chickens on a large scale. They are made in any lengths from 6 to 16 feet, to suit. One lamp is sufficient to heat a large 16-foot brooder, and this may be divided into three or four compartments, as desired, having a total brooding capacity of 300 to 400 chicks. It reduces the labor of handling a large flock of chickens to a minimum, only 20 minutes per day being required to attend to the brooder. The heat is uniform in every compartment. These are not pipe brooders, but furnished with our famous tank system, which supplies a uniform heat in every part of the brooding room, far better than any pipe system, which, although thoroughly tried by us, was found too unreliable when applied on a small scale, besides being more expensive. All these brooders can be had with copper tanks, which will outlast iron pipes, while the cheaper galvanized tanks, if always kept *full of water*, will last many years. You can brood chicks from one day old up to ten weeks of age, as the floor of each compartment can be adjusted, thus giving the small chick plenty of heat, and by dropping the floor for the older ones all ages can be brooded at the same time.

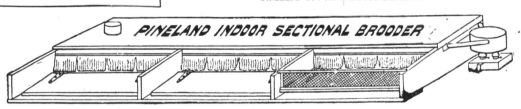

PINELAND INDOOR SECTIONAL BROODER

PRICES OF BROODERS.	Capacity.	Length.	Width.	Height.	Weight, Crated.	COPPER TANKS.		GALVANIZED TANKS.	
						Manufacturer's Price.	P. H. & Co.'s Net Price.	Manufacturer's Price.	P. H. & Co.'s Net Price.
Outdoor Brooder.............	50 Chicks.	3 ft.	4 ft.	3½ ft.	125 lbs.	$16.85	$16.00	$15.00	$14.25
" "	100 "	4 "	4 "	3½ "	175 "	22.50	20.25	20.00	18.00
" "	150 "	6 "	4 "	3½ "	225 "	28.50	25.65	25.00	22.50
" "	200 "	8 "	4 "	3½ "	300 "	34.75	31.28	30.00	27.00
Indoor Brooder, Junior......	100 "	5 "	2½ "	1 "	85 "	11.75	11.16	10.00	9.50
" "	150 "	5 "	3½ "	1 "	100 "	18.00	16.20	15.00	13.50
" " Sectional..	25 Chicks per ft.	6 to 11 ft.	2 "	1 "	25 lbs. per ft.	3.25 per ft.	2.93 per ft.	2.60 per ft.	2.34 per ft.
" " " ..	25 " "	12 " 16 "	2 "	1 "	25 " "	2.85 "	2.57 "	2.40 "	2.16 "

BONE FOODS FOR POULTRY.

For best success in poultry raising, and particularly for egg production, pure fresh bone foods are indispensable. Bone, meat and gristle supply the nitrogen, albumen and lime of which the white, yolk and shell of eggs are largely composed. Bone is a safe food—not stimulating—there is no injurious reaction. For fattening capons and market poultry it is economical and invaluable. It is, of course, essential for best results that bone foods be prepared while fresh and untainted. Fertilizer factory bone, which is often cut with acid, should never be used. The best method of supplying your poultry with bone is to get it fresh from your butcher as needed, and cut it with a bone cutter (such as we offer on page 27). The best substitutes for this are:

Henderson's Poultry Bone, crushed or ground from fresh raw knuckle bones, guaranteed untreated, which we offer in the following grades:

Henderson's Granulated Poultry Bone, crushed pieces, 1/16 to 1/8 inch in size. PRICES: 10c. per lb.; 3 lbs., 25c.; 25 lbs., $1.00; 100 lbs., $2.75.

Henderson's Ground Poultry Bone. Contains both fine and coarse particles. PRICES: 10c. per lb.; 3 lbs., 25c.; 25 lbs., $1.00; 100 lbs., $2.25.

Henderson's Poultry Bone Meal. Sifted down to meal and flour size; fine for mixing in the mash. PRICES: 10c. per lb.; 3 lbs., 25c.; 25 lbs., $1.00; 100 lbs., $2.25.

SMITH & ROMAIN'S PURE MEAT MEAL.

An extract of meat and bone. This is composed of fresh livers and lights of beef, cattle and sheep heads, thoroughly boiled, dried and ground to meal. It takes the place of insect life for poultry, contains the nutritious elements of meat, the egg productiveness of bone, adds gloss and beauty to plumage, makes chicks grow and fills the egg basket, prevents constipation, counteracts egg-eating and feather-pulling, and makes good the waste and drain of laying and molting hens. Special feeding directions by A. F. Hunter to each purchaser. PRICES: 10 lbs., 35c.; 25 lbs., 75c.; 100 lbs., $2.25.

SUPERIOR GRANULATED BEEF SCRAP.

A prepared meat food for poultry, guaranteed produced from sound meat only. One of the finest foods to make hens lay, and particularly valuable in the winter. Feed three mornings a week in the hot food a handful for every three fowls. PRICES: 5 lbs., 25c.; 25 lbs., $1.00; 100 lbs., $3.00.

FIDELITY FOOD FOR FOWLS.

One of the best foods. Largely used by practical growers, from whom many letters testifying to its superior merits have been received. It is fed dry, no mash mixing, no fermenting of food and consequent disease. It contains the carbohydrates and protein constituents and all food essentials in a well-balanced combination to promote health, thriftiness and egg production. Scatter the food through the straw, leaves or litter; the chickens will eagerly search and scratch for every granule of it. PRICES: 25-lb. bag, $1.25; 50 lbs., $2.00; $3.50 per 100 lbs.

FIDELITY FOOD FOR CHICKS.

A safe and nourishing food for young chicks just out of the shell, and can be depended upon to carry them right along until maturity. It is fed dry, though plenty of fresh water should be at hand for the little fellows to drink. Wet foods are unnatural for gallinaceous birds, and should never be employed except for fattening for market; besides, young chicks must be kept warm, and moist foods in a warm room soon turn sour, causing the chicks to dwindle away and die. **Fidelity Food** is the outcome of extensive experiments to overcome the above serious difficulty in the artificial raising of poultry, and it has long been used and recommended by the leading poultry raisers of the country. PRICES: 25c. per pkg.; 25 lbs., $1.25; 50 lbs., $2.00; 100 lbs., $3.50.

CHICK MANNA.

The first ten-day food for chicks just hatched. Promotes health, quick growth and strong development. Let it be the first food, and be exclusively fed for ten days or more, after which other food may be given which should contain a portion of Chick Manna in daily decreasing quantities for a few days more. Full directions on packages. PRICES: 1-lb. pkg., 10c., $1.00 per doz.; 5-lb. pkgs., 45c.; one case of ten 5-lb. pkgs. equals 50 lbs., for $4.00; 60-lb. case in bulk, $4.25.

"ROWEN," OR SECOND-CROP CLOVER HAY.

From the Woodhid Farm. Cured scientifically for feeding poultry as a green food during the winter or all the year round, when fowls are confined and do not have access to green grass. Of all the vegetable foods clover is the most valuable—it has great food value, containing both nitrogen and lime. It increases the egg yield, makes them richer and sweeter and of stronger fertility, dilutes and makes the food ration bulky—counteracting the effects of concentrated foods—makes them digestible, corrects constipation, tones up the health and prevents accumulation of internal fat. It may be used as a food by itself, or as the basis of the morning mash. Full directions for use by A. F. Hunter (editor *Farm Poultry*), to every purchaser. 50-lb. bag, $1.10; 100 lbs., $2.00; 500 lbs., $7.50.

CLOVER MEAL.

Pure clover hay ground to a fine meal, a practical form in which to feed it to fowls, as it is prepared by simply pouring enough hot water over it to make it crumbly, not sloppy. PRICES: 10-lb. pkg., 40c.; 25-lb. bag, 75c.; 50-lb. bag, $1.25; 100-lb. bag, $2.25.

VARIOUS GRAIN AND SEEDS FOR POULTRY AND BIRDS.

ALL FRESH, PURE AND SUPERIOR.

Canary, best Sicily	10c. per lb.;	$7.00	per 100 lbs.
Hemp	10c. "	6.00	"
Maw	20c. "	16.00	"
Millet	10c. "	3.00	"
German Rape	10c. "	7.00	"
Vetches for Pigeons	10c. "	6.00	"
Peas "	6c. "	3.00	"
Mixed Bird Seed for Canaries	10c. "	7.00	"
Sunflower	10c. "	6.50	"
Kaffir Corn	12c. "	6.00	"
Corn, Shelled		1.25	per bushel
Wheat		1.25	"

MASTICATORY ESSENTIALS for POULTRY.

The feathered tribes, having no teeth, can only masticate food by the aid of hard, sharp substances which act in their gizzards as grinders. Round, smooth-edged gravel, etc., while better than nothing, does not compare with hard, ragged-edged substances, not too small—even sharp bits of glass do no injury to the gizzard. Where poultry has the run of a farm, unless it is very sandy, they can usually find an ample supply; but where flocks are kept in limited space, particularly in poultry yards and houses, the supply of grinding material soon becomes exhausted, and must be furnished if success is to be attained. Hence we offer as the best thing for this purpose—

MANN'S GRANITE CRYSTAL GRIT.

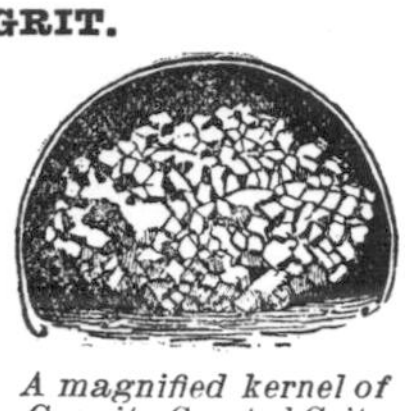

A magnified kernel of Granite Crystal Grit.

For chickens, ducks, turkeys, geese, pigeons and birds. It makes thorough digestion of food possible, and, in consequence, invigorates, improves the health, increases the number of eggs, aids in rapid gain in size and weight, matures broilers earlier, prevents sour crops and attendant diseases, and saves food by enabling poultry to utilize all that is eaten. **Granite Crystal Grit** is far superior to any other, clean and free from dust. It is the sharpest and hardest grit known; in every piece of grit as large as a kernel of corn are hundreds of sharp thousand-cornered crystals that grind and cut and never wear round and smooth. It is light in color, and, consequently, attractive to fowls. Granite Grit never powders like the "mica" and other grits from which cause hens often refuse them. **In ordering Granite Crystal Grit** please state size of Grit wanted: No. X for brooders and pigeons; No. XX for chickens; No. XXX for hens, ducks and turkeys. PRICES 25-lb. bag, 50c.; 50-lb. bag, 75c.; 100-lb. bag, $1.00.

PEERLESS BRAND OF CRUSHED OYSTER SHELLS.

An important article to keep before fowl, especially laying hens, throughout the season, and particularly so in the winter. While the sharp-edged pieces aid in grinding the food in the gizzard, yet a portion of the constituents of oyster shells is consumed as food, and is largely utilized in egg formation, making eggs larger and heavier; the lime in them prevents soft shells, and makes them strong enough to carry without breaking. This **Peerless Brand** we offer as a high-class article, not refuse; they are fresh dried by a patent hot-air process, not by direct fire, which burns out many good qualities; they are then crushed, and the dust and dirt blown out by a patent fan while being screened, leaving them pure and white. PRICES: 5-lb. carton, 10c. each; $1.00 per doz.; per 100-lb. bag, $1.00.

FRESH POULTRY CHARCOAL.

This is absolutely pure, triple burned, containing no dirt, and is made from selected wood. The benefits of using charcoal for poultry are very pronounced—it sweetens both food and crop, and corrects the bowels. The granulated is for use in open dishes, the powdered for mixing in the food mess. PRICES: **Granulated,** 5-lb. pkg., 40c.; 25 lbs., $1.75; 100 lbs., $5.50. **Powdered,** 5-lb. pkg., 35c.; 25 lbs., $1.60; 100 lbs., $5.00.

Medicinal Poultry Foods AND Specifics.

SHERIDAN'S CONDITION POWDER.

Prevents and cures diseases of hens—especially valuable in molting time. It is strictly a medicine, to be given in small doses once daily in their food. PRICES: Small pkg., 25c.; 5 for $1.00; large pkg., $1.00; 6 for $5.00.

PRATT'S POULTRY FOOD.

A cure and preventive of chicken cholera, roup, gapes and other diseases. It helps molting fowls and will make them lay sooner, and will increase the quantity of eggs. Young chicks grow healthy and quickly; turkeys, ducks, geese and pigeons thrive on it. PRICES: 26-oz. pkg., 25c.; 5-lb. pkg., 60c.; 12-lb. bag, $1.25.

AGATHA SPECIFIC FOR LAYING HENS.

A specially prepared food, promoting the health of laying hens, increases egg production, and supports them through molting. It is also valuable for all classes of poultry. 5-lb. pkg., 35c.

MADOC GAPE CURE.

Quickly effective and a sure cure for gapes. It soon causes the gape worms in the wind-pipe to lose their power, and the cause being removed the chicks soon regain vigor. It is used by mixing with the feed. PRICES: 6-oz. can, 20c.; 14-oz. can, 35c.

ROUP PREPARATION.

F. P. Cassel's celebrated specific, to be dissolved in boiling water and given in the drinking water, prevents the spreading of the disease and is a cure if the disease is not too advanced. If advanced, it is to be used in connection with the Multum in Parvo Poultry Powder offered below. Price Roup Preparation: ½-lb. pkg., 25c.; 1-lb. pkg., 40c.

MULTUM IN PARVO POULTRY POWDER.

A reliable cure for chicken and turkey cholera, and valuable in connection with Roup Preparation, offered above, as a remedy in advanced stages of roup. PRICES: ½-lb. pkg., 20c.; 1 lb., 35c.

CARY'S TURKEY PILLS.

Two pills, six to eight hours after turkeys are hatched, prevent colds and the numerous diseases so prevalent among the young turkeys; in metallic boxes, keeping indefinitely. Full directions and treatise on management of turkeys with each box. PRICE, 85c. per box.

LINSEED OIL MEAL.

Very beneficial if a little is mixed with the soft feed occasionally; causes rich, glossy plumage. 10 lbs., 50c.; 25 lbs., $1.00.

Tobacco Stems. A few in the nests and around the houses and coops prevent vermin. (See page 62.)

Tobacco Dust. Useful to dust poultry houses and coops. (See page 62.)

Persian Insect Powder. One of the best non-poisonous flea and insect destroyers. (See page 62.)

Slug Shot. A cheap and effectual non-poisonous insect powder. (See page 62.)

Land Plaster. Sprinkled around the poultry houses and yards adds to their attractiveness; is a good absorbent and counteracts bad odors. (See page 69.)

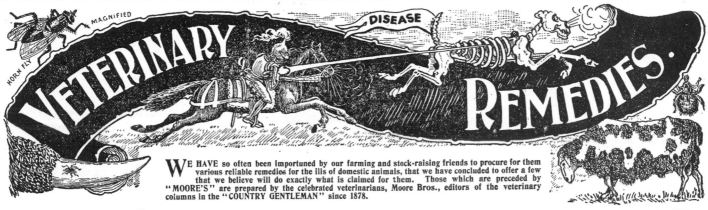

VETERINARY DISEASE REMEDIES.

WE HAVE so often been importuned by our farming and stock-raising friends to procure for them various reliable remedies for the ills of domestic animals, that we have concluded to offer a few that we believe will do exactly what is claimed for them. Those which are preceded by "MOORE'S" are prepared by the celebrated veterinarians, Moore Bros., editors of the veterinary columns in the "COUNTRY GENTLEMAN" since 1878.

HAMMOND'S "CATTLE COMFORT"
PROTECTS CATTLE, HORSES, MULES, DOGS, FOWLS, ETC., From Horn Fly, Gnats, Mosquitoes and other Insects.

Horn Flies settle at the base of the horns, they light on the body when feeding and insert their beaks into the skin, causing irritation and inflammation, making cows so restless as to lose flesh and shrink milk. Horses are disfigured and annoyed. In hot weather, apply "Cattle Comfort," mixed with equal quantity of kerosene, with a cloth by moistening the base of the horns, along the back to the root of the tail, and on the neck and fore quarters; or if the application includes feet, legs and rest of the body, it will drive every fly away. One application will last 10 days or more in dry weather.

The application of Cattle Comfort will also relieve Mules, Horses, Dogs and Fowls from the effects of flies and insects, and, besides, is healing to any sore. Applied to the perches in the hennery it prevents the spread of lice; put on the heads of fowls it destroys head lice; applied to many dogs it affords relief and effects a cure.
Prices: 1-quart can, 40c.; 1 gallon, $1.15; 5 gallons, $5.00.

"P. D. Q." POWDER for DOGS and POULTRY.

This celebrated, non-poisonous, disinfecting insect powder is considered the best by many prominent breeders and fanciers, who use it exclusively for fleas, lice, and other insects on dogs, cats, chickens, cattle, etc. It kills the insects immediately, and does not injure hair, feathers or skin—in fact, is healing and a disinfectant, and "slicks up the coat." Price, 1-lb. box, 25c.; 5-lb. bag, $1.00.

MOORE'S GENERAL COW DRINK.
For Cows, Yearlings and Calves.

This is an article that every dairyman, farmer and breeder of cattle ought to have on hand. As a preventive and cure for the under-mentioned diseases it has no equal, and will save many dollars and frequently the loss of a valuable animal, besides saves much anxiety and worry.

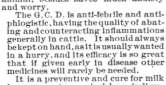

The G. C. D. is anti-febrile and anti-phlogistic, having the quality of abating and counteracting inflammations generally in cattle. It should always be kept on hand, as it is usually wanted in a hurry, and its efficacy is so great that if given early in disease other medicines will rarely be needed.

It is a preventive and cure for milk fever, garget or caked bag, indigestion, bloating and maw-bound, retention of afterbirth, fevers, unacclimated cattle, black leg, colic, diarrhœa, apoplexy, staggers, cowpox, sore mouth, red murrain or red water and the spread of the disease, sudden loss of milk or cud. It keeps your cattle in a healthy, flourishing condition and is used and recommended by some of the largest raisers of fine stock in this country.
Price, 50c. per can, $5.00 per dozen.

MOORE'S CLEANSING DRENCH. For removal of afterbirth or cleansing cows after calving. 50c.; $5.00 doz.
MOORE'S CATTLE TONIC. For cows that have been sick, or are run down from any cause, and need building up. 50c.; $5.00 doz.
MOORE'S MEDICINE FOR COWS GIVING BLOODY MILK. $2.00 per can.
MOORE'S INSTRUMENT FOR OPENING OBSTRUCTED TEATS. $1.00; **LEAD PROBE** for same. 25c.

ENGLISH "COLD WATER" SHEEP DIP.
"THYMO-CRESOL."

The safest, handiest, most effective and most economical Sheep Dip and Cattle Wash made. It is neither poisonous nor corrosive, safe to use in coldest weather, mixes instantly with cold water, promotes the growth and improves the quality of the wool; it leaves the fibre of the wool not only absolutely uninjured, but, from its oily nature, conserves and adds to the natural grease or yolk, and gives wool increased weight, lustre, softness and brilliancy, so that where this dip is used, other things being equal, the top ruling price may be obtained for the clip. For most purposes a gallon of it makes as much as 100 gallons for actual use. It is free from poisonous ingredients and harmless to man and domestic animals; in fact its antiseptic nature effects a rapid healing.

And yet it is quick death to insects; it does not poison them, but dries them sufficiently to close their breathing pores in their skin, thus suffocating them.

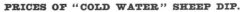

It is also a quick, safe and sure cure for scab, ticks, red lice, fly, maggots and foot rot in sheep; of inestimable value wherever stock is grown, both as an external wash and internal drink, for cattle, horses, dogs, pigs and poultry; kills lice, nits and other insects; cures thrush, grease cracked heels, mange, canker and skin diseases; ring worm, wounds, sores, saddle galls, burns, scalds, etc.; "worm in the throat" of lambs, gapes in chickens, tape worm in calves, etc. Of great value mixed in whitewash for chicken houses, scaly trees, etc.

This English Cold Water Dip has received numbers of gold and silver medals and diplomas in Europe, including the highest award at our World's Fair in Chicago. It is used and recommended by prominent veterinarians, breeders and farmers on both sides of the water. Among those in the United States are Mr. Henry Stewart, author of the "Shepherd's Manual"; Wilmer Atkinson, editor of the "Farm Journal"; Frederick Bronson, the famous whip, and many others, whose testimony is printed in the special "Sheep Dip Circular," which we will mail on application. Sample 2-oz. bottle, including mailing case and postage, 25c.

PRICES OF "COLD WATER" SHEEP DIP.

4-oz. bottle, 20c.; 8-oz. bottle, 35c.; pint can, 45c.; quart can, 65c.; half-gallon can, $1.00; 1-gallon can, $1.75; 2-gallon can, $3.25; 3-gallon can, $4.50; 5-gallon can, $6.75; 10-gallon can, $12.00.

MOORE'S GALL POWDER will improve, heal and toughen immediately and cure speedily, even while the animal works, collar galls, bit galls, saddle galls and any skin abrasions on horses or mules. 50c. and $1.00 packages.
MOORE'S GOLDEN BLISTER. For hip and shoulder lameness, sprained tendons, spavins, splints, contracted or brittle hoofs, and cracks, callouses, puffs, etc., on horses; leaves neither blemish nor scar. It also quickly cures ring worm on cattle. $1.00.
MOORE'S WORM AND TONIC POWDER. For sheep, horses, mules, cattle, dogs and swine. It is death to worms and absolutely harmless to animals, on which it acts as a tonic. Worms kill more sheep than dogs. $1.00 per can. (Cheaper in bulk to feed flocks as a preventive.)
MOORE'S COLIC MEDICINE. $1.00 per bottle.
MOORE'S PHYSIC. Fever, Cough, Diuretic, Alterative and Worm Balls. 50c. each.

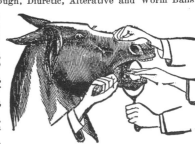

MOORE'S FOUL HOOF AND FOOT ROT OINTMENT. A certain and speedy cure for "foul hoof" in cattle and "foot rot" in sheep. $1.00.
MOORE'S OINTMENT FOR SCRATCHES. 50c. and $1.00 per box.
MOORE'S OINTMENT FOR PROUD FLESH. $1.00 per box.
MOORE'S LINIMENT FOR CAPPED HOCKS. $1.00.
MOORE'S BALLING IRON. For holding animal's mouth open. (See cut.) $1.00.
MOORE'S SAFETY MILKING INSTRUMENT. See page 44.

Full directions accompany every article. All medicines will keep good for an indefinite period.

REMEDIES for

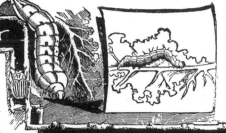

WE give below a list of vegetables, fruits, trees and plants, and in italics the insects or diseases that usually attack them, followed by one or more numbers. To find out what remedy to use, refer to a corresponding number under the headings of "Remedies for Insects," or "Remedies for Fungus." For instance, if an insect is attacking your Asparagus, by referring to the list of vegetables you will find that it is probably the beetle that is causing the devastation; the numbers following Asparagus beetle are 9 and 13. By reference to the corresponding numbers under "Remedies for Insects" you will find that Paris green in solution or Slug Shot blown on in powder form are the remedies to apply.

INSECTS & FUNGUS

For Insects, etc., troubling Garden Vegetables.

Use for Asparagus, *beetle,* 9 or 13; Bean, *rust,* 23, 24. Cabbage and Cauliflower, *worm or slug,* 6, 11, 13; *cutworm,* 10; *flea beetle,* 8, 14, 16, 13; *maggot,* 15, 17. Carrot, *worm,* 9, 13. Celery, *worm,* 9, 13, 11; *rust,* 21, 24. Corn, *cutworm,* 10; *borer,* 10; *worm,* 6, 13. Cucumber, *worm,* 6, 13; *beetles,* 8, 13. Egg Plant, *potato bug,* 8, 9, 13. Endive and Lettuce, *worm,* 11, 13. Melon, Squash and Pumpkin, *worm,* 6, 13; *bugs and beetles,* 8, 13; *root borer,* 15; *leaf rust,* 24, 23. Onion, *maggot,* 15, 17. Parsley and Parsnip, *worm,* 8, 9, 13. Potato, *bug,* 8, 9, 13; *blight,* 23, 24.

Radish, *flea beetle,* 8, 14, 16, 13; *maggot,* 15, 17. Sweet Potato, *sawfly worm,* 6, 8, 11, 13. Tomato, *worm,* 6, 13. Turnip, *flea beetle,* 13, 16, 8; *maggot,* 15, 17.

Small Fruits. Cranberry, *worm,* 8, 9. Currants, Gooseberries, Raspberries, *worms,* 6, 7, 8, 13; *leaf hoppers,* 11, 13; *rust and mildew,* 21, 22, 23, 24. Grapes, *slug,* 9, 8, 7, 6, 13; *flea beetle,* 9, 8, 13, 16; *thrip,* 14; *rose chafer,* 11, 13; *black rot, scab, mildew and rust,* 21, 23, 24. Strawberry, *slugs and worms,* 6, 7, 13; *leaf roller,* 6, 7; *root louse,* 15, 17.

Fruit & Ornamental Trees & Shrubs. Bark louse, 3, 4, 5, 20; *caterpillar, slugs, worms,* 2, 9, 13;

beetles, aphis, 14, 3, 4; *curculio,* 9; *rose chafer,* 11; *borers,* 2; *mildew and leaf rust,* 22, 23, 24; *scale,* 2, 3, 4, 5, 14.

Flowering and Ornamental Plants. Aphis or green fly, 11, 18, 16, 14, 13; *worms, slugs and caterpillars,* 11, 9, 8, 6, 13; *rosebug or chafer,* 11, 13; *leaf hoppers,* 19, 13. 14; *red spider,* 3, 4, 14; *mealy bug and scale,* 3, 4, 5. 14; *blue root louse,* 15, 17; *thrip,* 3, 4, 14; *leaf rust and mildew,* 21, 22, 23, 24. Hollyhock Disease, 21, 24.

Cotton. Worm, 9, 13.

Lawns. Ants, 1.

⬥ REMEDIES FOR INSECTS. ✠ ⬥

Imported German Caterpillar Lime.

A highly recommended European remedy for the prevention of crawling insects from going up or down the trunks of trees; prevents the laying of eggs on the bark, and the hatching of those already laid; prevents borers, scale, etc., and keeps all animals from gnawing the trees; it remains sticky and efficacious from 3 to 5 months, and is easily and cheaply applied. (*Little booklet, giving full information, mailed on application.*) Price: 5 lb. can, $1.00; 10 lb. can, $1.75; 25 lb. keg, $3.75; 50 lb. keg, $6.75; 100 lb. keg, $12.75.

No. 2. *Apply with a paddle and spread evenly with a hard brush, leaving the lime ¼ inch thick. For crawling insects, a ring of it 4 inches wide around the tree will suffice. Complete instructions with each package.*

FIR TREE OIL.

The most popular, and probably for "all round" purposes, the best insecticide in the market for greenhouse and house plants; it frees plants of nearly all insects to which they are subject, and for the following it has no superior: Mealy bug, scale, red spider, aphis (black and green), thrip, blight, worms and slugs, and is also a valuable remedy for animal parasites and insects. Price, ½ pint tin, 40c.; pint, 75c.; qt., $1.40; ½ gal., $2.50; gal., $4 25.

No. 3. *Dilute ½ pint to about 10 gallons of soft or rain water (for tender plants or young growth make weaker). Spray on with vaporizer, syringe or bellows.*

FIR TREE OIL SOAP.

A formula of the above Fir Tree Oil, prepared in soap form and preferred by some. It will do all claimed for the fluid preparation. Price, ½ lb. tins, 25c.; 2 lbs., 75c.; 5 lbs., $1.75; 10 lbs., $3.25; 20 lbs., $6.00.

No. 4. *One ounce makes a gallon of non-poisonous insect-killing liquid. Apply with syringe or bellows.*

KEROSENE EMULSION—(A Paste).

For plant lice of any kind, cabbage worm, scale insects on apple, pear, orange, lemon and other trees. Price, 1 lb., 2cc.; 5 lbs., 75c.; 25 lbs., $2.50.

No. 5. *Mix 1 lb. Emulsion to 10 or 15 gallons water, according to strength required. Spray on with syringe or force pump.*

HELLEBORE, POWDERED WHITE.

For the destruction of slugs, worms, caterpillars, etc. Less poisonous than Paris green and London purple, and safer to use when fruits or vegetables are nearly ripe. Price, 20c. per lb.

No. 6. Dry Application. *Dust on dry or mix 1 lb. with 4 lbs. of powdered lime or flour; dust on with powder duster or bellows or gun.*

No. 7. In Solution. *Dissolve 1 oz. to 3 gallons of water; apply with syringe, pump or vaporizer.*

PARIS GREEN.

A poisonous, insoluble powder, indispensable on the farm or garden—for preventing the ravages of potato bugs, codling moth, worms, caterpillars, slugs and bugs. Price, 25c. per lb.

No. 8. Dry Application. *Mix with plaster, flour or other dilutant—1 part to 100; apply with duster, bellows or gun.*

No. 9. In Solution. *Mix 1 lb. to 200 or 300 gallons of water. Apply with pump, syringe or vaporizer.*

No. 10. Bait. *For cut worms, mix 1 oz. with 8 ozs. of syrup; mix thoroughly with fresh chopped grass or leaves.*

PERSIAN POWDER OR BUHACH.
Henderson's Superior Grade.

Cheap adulterated grades are worthless. A most effective *non-poisonous* impalpable powder—so fine that it penetrates the innermost crevices—for worms, flies, aphis, and almost all kinds of insects—it is very effectual—suffocating them by filling up the breathing pores. Price, 10c. ¼ lb.; 35c. lb.

No. 11. Dry. *Blow on with a bellows or gun.*

No. 12. Solution. *Dissolve 1 oz. to 3 gallons of water, and apply with syringe or pump.*

SLUG SHOT.

A non-poisonous powder and a very popular insecticide —it requires no further mixing or preparation—easily applied and not injurious or dangerous to animals, the person applying it, or fruits and vegetables treated. Very effectual in destroying potato bugs and beetles, green and black fly, slugs, worms, caterpillars, etc. (*By express or freight only.*) Price, per keg of 125 lbs. net, $5.00; per barrel of 235 lbs. net, $8.50; in 10 lb. packages. each, 50c.; 5 lb. packages, 30c.; in tin canister, with perforated top for applying, 25c. each.

No. 13. *Apply with duster, bellows or gun. 10 to 40 lbs. is sufficient for an acre.*

ROSE LEAF INSECTICIDE.
Extract of Tobacco.

One of the most effectual articles for the destruction of all insects and scale on plants; a pure concentrated extract of tobacco. Full instructions for the various methods of applying, including testimonials from leading florists, are given in our "Rose Leaf" pamphlet, mailed on application. Pint can, 30c.; quart can, 55c.; gallon can, $1.50; 5 gallon can, $5.50.

No. 14. *Dilute with from 30 to 150 parts of water, according to the delicacy of the plants to be treated, and apply with spraying bellows or syringe with spraying nozzle.*

No. 15. *Dilute as above, and apply with watering pot sufficient to reach the roots.*

TOBACCO DUST, FINE.

For green and black aphis. fleas, beetles, etc. Splendid fertilizer and preventtive for insects in the ground and around roots. Price, 10c. per lb.; 5 lbs., 30c.; 10 lbs., 50c.; $2.00 per 100 lb. bag.

No. 16. *For insects on plants, apply with powder duster or bellows.*

No. 17. *For worms or grubs in the soil, apply liberally to the surface and rake in, or strew thickly in the drills before planting.*

TOBACCO STEMS.

Indispensable for fumigating greenhouses and conservatories—for the destruction of green and black aphis, and other insects. Price, 50 lb. bale, 85c.; 100 lb., $1.50; per ton 2,000 lbs., $20.00.

No. 18. *Dampen thoroughly a few hours before using, place about a half pound over a handful of shavings in a fumigator, and light.*

WHALE OIL SOAP.

Makes an excellent wash for trees and plants where insects and eggs affect the bark, and for smearing on the trunks of trees to prevent worms from crawling up. Price, 15c. per lb.; 2 lbs., 25c.; 5 lbs., 50c.; 25 lbs. and over, at 8c. per lb.

No. 19. *For insects on plants, dissolve ¼ lb. to a gallon of water; apply with syringe or spraying pump.*

No. 20. *For bark insects, etc., dilute with water to consistency of thick paint, and apply with brush.*

ANT DESTROYER.

A non-poisonous powder which will destroy or drive away black ants from lawns, trees, plants, houses or other affected locality. Price, ¼ lb. can, 25c., 1 lb. can, 75c.

No. 1. *Sprinkle it around near the haunts.*

REMEDIES FOR FUNGUS, SUCH AS MILDEW, RUST, BLACK ROT, ETC.

COPPER SOLUTION.
AMMONIATED.

A Fungicide, the same as Bordeaux Mixture, the essential ingredient, "*Carbonate of Copper*," being dissolved in ammonia in this, while in Bordeaux it is counteracted by lime. Bordeaux is the cheaper and most popular for all ordinary purposes, but for late sprayings, when fruits are nearing maturity, or plants in bloom, Copper Solution is usually used, as there is no lime or limy sediment left to be washed off by rain or hand before marketing. Price, 1 qt., 50c.; 1 gal., $1.50; 5 gals., $6.00.

No. 21. *Dilute 1 quart to 25 gallons of water; apply with Knapsack sprayer or bellows vaporizer.*

FOSTITE.

A famous French preparation, containing silicate of magnesia and copper sulphate, powdered so minutely that when puffed or blown from a powder-gun or bellows, it forms a cloud which settles evenly over foliage and plants, and is not only of inestimable value for all fungoid diseases—mildew, black rot, rust, leaf blight, etc.—but it is also a splendid insect destroyer. It is a powder. Price, 5 lb. package, 50c.; 25 lb. box, $2.00; 50 lb. box, $3.50; 100 lb. box, $6.50.

No. 23. *Blow on with a power-bellows or powder-gun, while the foliage is moist, either with dew or after syringing. At the rate of 25 lbs. per acre for grapevines, for each application.*

BORDEAUX MIXTURE, IMPROVED.

Ready for use by simply adding cold water. An indispensable Fungicide, curing and preventing black rot, mildew, blight, rust, scab and all fungoid diseases on fruits and plants. (1 lb. makes 5 gallons of spray.) Price, 1 lb., 20c.; 5 lbs., 75c.; 10 lbs., $1.25; 25 lbs., $2.50; 100 lbs. or over, at 9c. lb. extra.

No. 24. *Dissolve 1 pound to 5 gallons of water, and apply with Knapsack sprayer or vaporizer, or pump with vaporizing nozzle.*

FLOWERS OF SULPHUR.

Price. 10c. lb.; 10 lbs., 60c.; 25 lbs. and over, at 5c. lb.

No. 22. *Apply with bellows or gun.*

· · · SPRAYING CALENDAR. · · ·

In the preparation of this calendar the most important points regarding sprays have been selected and arranged in such a manner that the grower can see at a glance what to apply and when to make the applications. Only the more important insect and fungus enemies are mentioned, though other enemies are also kept under control. Some applications are italicized, and these are the ones which are *most important*. The numbers refer to the insecticide or fungicide preparation described on the opposite page. Thus, you will see No. 9 is Paris Green in solution; No. 24, Bordeaux Mixture, etc.

PLANT.	First Application.	Second Application.	Third Application.	Fourth Application.	Fifth Application.	Sixth Application.
APPLE (Scab, codlin moth, bud moth.)	When buds are swelling, use No. 21.	*4–7 days before blossoms open, No. 24. For bud moth, No. 9, when leaf buds open.*	*When blossoms have fallen, Nos. 24 and 9.*	*8–12 days later, Nos. 24 and 9.*	10–14 days later, No. 24.	10–14 days later, No. 24.
BEAN (Anthracnose.)	*When third leaf expands, No. 24.*	*10 days later, No. 24.*	14 days later, No. 24.	14 days later, No. 24.		
CABBAGE (Worms, aphis.)	When worms or aphis are first seen, No. 5.	7–10 days later, if not heading, No. 5 or 7	*7–10 days later, if heading, hot water 130° F. or No. 7.*	Repeat third in 10–14 days if necessary.	(When plants are small, No. 9 may be used to check worms.)	
CHERRY (Rot, aphis slug.)	As buds are breaking, No. 24; when aphis appears, No. 5.	When fruit has set, No. 24. If slugs appear, No. 7 or 13.	10–14 days if rot appears, No. 24.	10–14 days later, No. 21.		
CURRANT (Mildew, worms.)	At first sign of worms, No. 9 or 7.	10 days later, No. 7. If leaves mildew, No. 24.	If worms persist, No. 7.			
GOOSEBERRY (Mildew, worms.)	When leaves expand, No. 24. For worms, No. 9 or 7.	10–14 days later, No. 24. For worms, No. 7.	10–14 days later, No. 21. For worms, No. 7.	10–14 days later, No. 21.		
GRAPE (Fungus diseases, flea beetle.)	In Spring, when buds swell, No. 21. For flea beetle, No. 9.	When leaves are 1–1½ inches in diameter, No. 24. For larvæ of flea beetle, No. 9.	*When flowers have fallen, No. 24.*	10–14 days later, No. 24.	10–14 days later, if any disease appears, No. 24.	10–14 days, No. 21. Later applications if necessary.
PEACH, NECTARINE, APRICOT (Rot, mildew.)	*Before buds swell, No. 21.*	Before flowers open, No. 24.	*When fruit has set, No. 24.*	*When fruit is nearly grown, No. 21.*	5–10 days later, repeat fourth.	5–10 days later, repeat fourth if necessary.
PEAR (Leaf blight, scab, psylla, codlin moth.)	As buds are swelling, No. 21.	*Just before blossoms open, No. 24. When leaves open for psylla, No. 5.*	After blossoms have fallen, Nos. 24 and 9.	8–12 days later, repeat third.	10–14 days later, No. 24. Apply forcibly for psylla, No. 5.	10–14 days later, repeat fifth if necessary.
PLUM (Fungus diseases, curculio.)	*During first warm days of early spring, No. 24 for black knot. When leaves are off in the fall, No. 5 for plum scale.*	*When buds are swelling, No. 24 for black knot and other fungous diseases. During mid-winter, No. 5 for plum scale.*	*When blossoms have fallen, No. 24. Begin to jar trees for curculio. Before buds start in spring, No. 5 for plum scale.*	10–14 days later, No. 24. Jar trees for curculio every 2–4 days. For San José scale, No. 5 when young appear in spring and summer.	10–20 days later, No. 24 for black knot. Jar trees for curculio. When young plum scale insects first appear in summer, No. 5.	10–20 days later, No. 24 for black knot. Later applications may be necessary to prevent leaf spot and fruit rot. Use No. 21.
POTATO (Scab, blight, beetles.)	Soak seed for scab in corrosive sublimate solution (2 ozs. to 16 gals. of water) for 90 minutes.	When beetles first appear, No. 9.	When vines are two-thirds grown, No. 24. For beetles, if necessary, No. 9.	10–15 days later, repeat third.	10–15 days later, No. 24 if necessary.	
QUINCE (Leaf and fruit spot.)	When blossom buds appear, No. 24.	When fruit has set, Nos. 24 and 9.	*10–20 days later, No. 24.*	*10–20 days later, No. 24.*	*10–20 days later, No. 24.*	
RASPBERRY, BLACKBERRY, DEWBERRY (Anthracnose, rust.)	Before buds break, No. 21. Cut out badly diseased canes.	During summer, if rust appears on leaves, No. 24.	Repeat second if necessary.	(Orange or red rust is treated best by destroying entirely the affected plants.)		
STRAWBERRY (Rust.)	*As first fruits are setting, No. 24.*	As first fruits are ripening, No. 21.	When last fruits are harvested, No. 24.	Repeat third if foliage rusts.	Repeat third if necessary.	(Young plants not in bearing may be treated throughout the fruiting season.)
TOMATO (Rot, blight.)	At first appearance of blight or rot under glass or out-of-doors, No. 24.	Repeat first if diseases are not checked.	Repeat first when necessary.			

IMPLEMENTS for APPLYING INSECTICIDES & FUNGICIDES

Treated   Untreated

THE "LENOX" SPRINKLER.

THE "SUCCESS" KNAPSACK SPRAYER.

A Knapsack Sprayer and Bucket Sprayer Combined.

Five-Gallon Copper Tank. Brass Pump with Bronze Ball Valves.

THE "SUCCESS" KNAPSACK SPRAYER.

THIS valuable arrangement, originally designed by the U. S. Agricultural Department, is used for applying fungicides, such as Bordeaux Mixtures, Ammoniacal Compound of Copper, and other fluid remedies in a mist-like spray, for the treatment of grapes and other vegetation, for the prevention and cure of mildew, black rot and kindred diseases, as well as for potato blight, etc.; with it a man can spray five to six acres of vines in a day. The machines are made entirely of copper and brass, and the chemicals will not corrode or rust them. The air chamber keeps up a steady pressure, so a continuous discharge is given. No grape grower can afford to be without one. The **New Improvements** recently made to our Success Knapsack Sprayer entirely obviate all objections to the old patterns. The **New Improvements** enable this sprayer to be used either as a knapsack or bucket sprayer, the illustration as here given showing its use as a knapsack. To be used as a bucket sprayer the handle and lever are removed and the extra handle with which the outfit is provided is placed in position as shown by the dotted lines. The handle for carrying it is shown in the cut. The straps used on our knapsack are made extra wide just where the weight comes upon the shoulders. They are *provided with a mechanical agitator.* **The Pump** may be worked with either the right or left hand. The drip cup is just below the air chamber, and is made extra wide, so that any leakage around the plunger is returned into the tank and not allowed to run down the operator's back. The attachment shown at A is for underspraying. **Price,** complete, as shown in cut, **$10.00.**

THE "LENOX" SPRINKLER.

A cheap sprinkler for applying poisonous fluids, such as Paris green water, etc., to grape vines, low trees, bushes, potatoes, garden vegetables, etc. Can be carried either on the back or by hand, as desired. It is made of galvanized iron, holds five gallons of fluid, which flows through the rubber tube and is forced out in a spray by pressing the rubber bulb held in the hand; this spray will reach 10 to 12 feet. **Price,** with single sprinkling attachment as shown in the cut, **$3.50;** or with double sprinkling attachment for two rows at once, **$5.00.**

The "Lenox" Sprayer is not adapted for throwing fungicides, such as Bordeaux Mixture, as the spray is not fine enough nor continuous as in the Knapsack Sprayer offered above.

KEROSENE SPRAYERS,

FOR MECHANICALLY MIXING KEROSENE AND WATER, DISPENSING WITH EMULSION.

Kerosene properly diluted with water has long been known as a splendid insecticide for bark insects, such as scale, mealy bug, etc., as well as for melon louse, plant lice, and all insects that suck the juices of plants or blood of animals. It is also repellent to most beetles and bugs; but the preparing and applying of kerosene as it had to be done (in an emulsion with soap, etc.), so that it would remain thoroughly mixed throughout, was tedious, and not always satisfactory; but with our new **Kerosene and Water Automatic Mixing Sprayers and Pumps** the oil and water are thoroughly mixed in any desired proportions while pumping. An indicator with gauge plate shows the proportions of oil and water to apply for various purposes.

THE KEROSENE KNAPSACK SPRAYER.

This apparatus is practically the same as the Knapsack Sprayer offered above, with the addition of a separate tank for holding kerosene, which is automatically mixed with water when the handle is operated. The proportions are regulated by an indicator. For field and garden crops, low trees and grape vines it is, perhaps, the most convenient and rapid style of sprayer.

The **Kerosene Tank** is readily detached from the main tank, and a cap is furnished with which to close the oil inlet. The outfit then becomes the Success Knapsack Sprayer offered above. **Price,** complete with undersprayer and agitator, **$14.00.**

The "PEERLESS"

KEROSENE SPRAYER AND AUTOMATIC MIXER.

The "Peerless" is intended for tree and orchard work; it mechanically mixes in the most thorough manner, and in any proportions desired, water and kerosene oil, and sprays it at the same time. Full directions how to operate and the proportions to mix for different purposes sent with each pump. The kerosene attachment can be removed, so the pump may be used as an ordinary spray pump. (*Prices do not include barrel.*)

"Peerless" Outfit No. "A." Pump, copper kerosene tank, with 12½ ft. of hose and connections for pole, with spray nozzle.................................**$16.00**

or, with two 12½ ft. lengths of hose, 2 pole connections and 2 nozzles.........................**18.50.**

For all brass chamber pump, add $2.50 extra.

THE "PEERLESS" KEROSENE MIXER AND SPRAYER.

The "BUCKET" KEROSENE and WATER SPRAYER

And AUTOMATIC MIXER,

WITH

REMOVABLE KEROSENE TANK.

This Bucket Kerosene Sprayer makes a most convenient article for general use in spraying shade trees, plants, poultry houses, etc. The apparatus can be used as a regular bucket pump in the usual way by detaching the kerosene tank. All working parts are brass, the oil tank being made of copper. Complete directions are furnished with each sprayer. **Price,** complete, as shown in the cut, with 4 feet of hose and "Bordeaux" Nozzle, **$6.50;** or, with a 7-feet section of hose and connection for a pole for tree spraying, **$8.00.**

THE "BUCKET" KEROSENE MIXER AND SPRAYER.

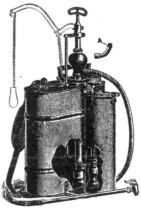

KEROSENE KNAPSACK SPRAYER.

"FRUITALL" SPRAYING OUTFIT.

A Strictly HIGH GRADE UP-TO-DATE PUMP, Embodying Several Important Improvements.

It is simple, the parts accessible and easily removed for cleaning if occasion demands. It has **all working parts of brass**. It can be mounted on or removed from barrel or tank by the *simple adjustment of two thumbscrews*. It is held firmly by clamp at top, and is adjustable to large or small barrel. The Brass Plunger is very strong. The Brass Valves never clog. Steel air chamber of unusually large capacity. It has a long adjustable lever. It is built low down, and with the exception of lever and discharge, is contained within the barrel, offering no obstruction to overhanging limbs, and is not top-heavy. The pump is large, but easily operated by a single person. **The Agitator is** mechanical in its action and much superior to the old method of "return discharge." This pump is the most efficient and satisfactory pump ever offered.

"**Fruitall**" Outfit "**A**," for single spray; Fruitall Spray Pump with one lead 10 feet, ½ in. discharge hose, with "Vermorel" Spray Nozzle and Agitator, **$8.50**; or, fitted to barrel, **$10.00**.

Iron Extension Pipe, 8 ft., with stopcock, to reach top of trees, fitted to attach to nozzle and hose, each **$1.25**.

GEM FORCE PUMP.

"FRUITALL" SPRAY PUMP.

The "STANDARD"
Double-Acting Spray Pump.

(Illustrated below, and also shown in operation on the back cover.)

A strong double-acting Spray Pump, with brass-lined cylinder and brass-cased differential plunger with brass valve seat, especially adapting it for diffusing poisonous mixtures, such as Bordeaux, Paris Green, Copper Sulphate, Kerosene Emulsions, etc., upon the trees, vines and bushes. The differential plunger forms the air chamber and contributes to sustaining a continuous and uniform discharge at spray nozzles, and besides it has an additional air chamber on the spout, which is of great advantage in old orchards, as the spray can be discharged to the topmost branches. The base is adapted for either end or side of barrel. Lever is extra long and strong. *Prices do not include barrel.*

PRICES:

Standard Outfit "A," for Single Spray: 2½ in. pump, with 2½ ft. suction pipe, brass strainer, and one 10 ft. lead of ½-in. hose with spray nozzle, **$12.50**; or, fitted to barrel, **$14.00**.

Standard Outfit "D," for Double Spray: Fitted for and supplied with two leads of discharge hose and two spray nozzles, **$15.75**; or, fitted to barrel, **$17.25**.

Extra iron extension pipes, 8 ft., with stopcock, base fitted to connect on hose; top threaded for spray nozzle, $1.25 each, extra.

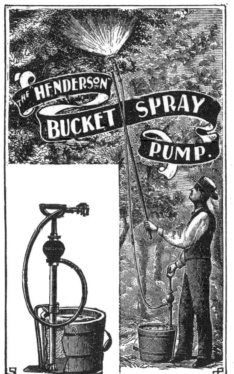

"HENDERSON" BUCKET PUMP, WITH POLE CONNECTION.

THE "SENTINEL"
Double-Acting Spray Pump.

(Illustrated on the right and also shown in operation on the back cover.)

Our "Sentinel" Double-Acting Spray Pump has brass-lined cylinder, brass valves and seats, plunger rod, etc., rendering it unaffected by acids, while the extra large air chamber especially adapts it for spraying tall trees. The suction and discharge valves are all grouped in valve chest, and are readily accessible by removing air chamber. This pump supplies the demand for a powerful Spray Pump of sufficient capacity to supply, if necessary, four leads of discharge hose for large orchards or groves. Cylinder 3 in. diam., double suction 1¼ in. for pipe. Air chamber 21½ inches high.

(Prices do not include tank or barrel.)

PRICES:

Sentinel Outfit "A." Pump with 5 ft. 1¼ in. rubber suction hose, and 2 leads, 25 ft. each, of ½-inch discharge hose, with 2 spray nozzles, **$41.00**.

Sentinel Outfit "B" is the same as the above, but fitted for and supplied with 4 leads of hose and 4 nozzles, **$55.00**.

Bamboo Extension Rods, for reaching to the tops of tall trees, fitted with ¼-inch pipe inside, shut-off cock, fitted for hose at one end and nozzle at the other, **$3.50** each, extra.

Gem Force Pump and Spraying Outfit.

Clamps to the chine or side of any barrel, but is made particularly for our Water Barrel and Truck outfits, offered on page 52. The "Gem" is intended to fill a want for a low-priced pump. It has not the power or capacity of the larger and more expensive spray pumps, but where there are only a few trees, bushes, vines, etc., to be treated, it will answer admirably for both applying insecticides and fungicides, and by altering the nozzle it is adapted for sprinkling walks, watering flower-beds, washing windows, carriages, etc. The pump can be quickly removed from the barrel.

PRICES
(without barrel or truck):

Gem Outfit "A." Pump fitted with two feet of suction pipe and brass strainer; 5½ ft. ½-inch hose and nozzle, **$5.00**.

Gem Outfit "B" is the same as "A," excepting the hose is 12½ feet long, fitted with pole connection to spray trees, **$6.00**.

For prices with barrel and truck, see page 52.

The "Henderson" Hand Bucket Pump.

This is the best and most durable portable hand pump in the market. Working parts are made of solid brass with large air chamber, and is double acting, throwing an absolutely continuous stream, either solid or in a fine spray, as desired; very light and easily carried, and works from any bucket or tub. Just the thing for throwing liquid insecticides and fungicides on low trees, shrubs, plants, etc., and by altering the nozzle it can be used for washing carriages, windows, etc. *(See cut.)* **Price, $3.50**; or, with extra 7 feet section of hose and pole connection for tree spraying, **$4.50**.

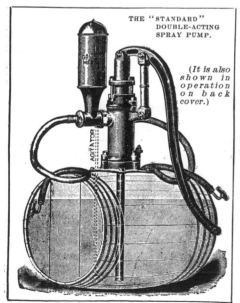

THE "STANDARD" DOUBLE-ACTING SPRAY PUMP.

(It is also shown in operation on back cover.)

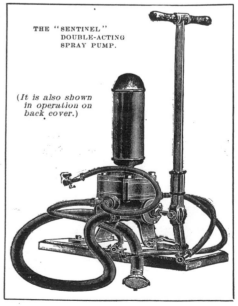

THE "SENTINEL" DOUBLE-ACTING SPRAY PUMP.

(It is also shown in operation on back cover.)

THE "SIROCCO"
HORSE POWER
POWDER DUSTER.

THE SIROCCO HORSE POWER POWDER DUSTER.

For rapidly dusting truckers' and farmers' crops of potatoes, cabbages, melons, grape vines, etc., with air-slacked lime, tobacco dust, plaster, slug shot, sulphur, dry soot, ashes, or any fungus or insect-killing powders. It does the work much more thoroughly, rapidly and with less waste than by any other implement; it can be regulated to cover the entire surface with the least possible quantity; the powder is so finely and evenly distributed that it floats in the air and settles down and envelopes every leaf and stem, thus smothering all bugs, caterpillars and slugs, which breathe through the pores of their skin. There is no excuse now for allowing your crops to be eaten up by insects, etc., on account of lack of time, for you can cover an acre with this machine in half an hour or less. The driver and horse are in front and are not affected, for they are driving away from the powder, which appears like a fog behind them. The machine is instantly thrown in or out of gear by a lever, and the funnels can be shifted to throw up in orchard trees or horizontally for field crops. The funnel on either side can be shut off if occasion requires.

Prices : 1-horse machine (*see cut*), tracks 2 ft. 4 in. wide, **$60.00.**
2 " " (straddles rows), " 3 " 7 " **75.00.**

THE "ELECTRIC" POTATO BUG EXTERMINATOR.

This is a cheap bellows, of large capacity for the price; it holds 1 pound of Paris green or other powder, which is blown and evenly distributed over the plants in large or small quantities, as desired, there being a perforated regulating tube inside which prevents any surplus being discharged. With it a rapid operator can treat an acre of potatoes in an hour. **Price, $1.00.**

HENDERSON'S VAPORIZING SYRINGE
FOR SPRAYING FLUIDS.

A low-priced zinc arrangement, just what is needed for small applications of fluid insecticides or fungicides, in a mist-like vapor; it holds 1 pint; length of barrel, 19 ins.; diam., 1¾ ins. Each, **$1.00.**

THE "RED JACKET" WHEELBARROW SPRAYER.
For applying Fluid Remedies for Insects or Fungus on all Garden Crops, Potatoes, Grape Vines, etc.

"You simply wheel the barrow, it does the rest." You can go over an acre of potatoes with it in 30 minutes.

It is a Single Wheel Automatic Sprayer, with tubular iron frame, and a wooden tank holding ten gallons. It is fitted with an endless sprocket chain gearing, and has a mechanical automatic agitator which travels in a semi-circle in bottom of tank and keeps the poison from settling. The pump and air chamber are brass, and the valves are bronze metal. It has double tube sprayers for two rows of potatoes, and has two Vermorel Nozzles. The tubes are so arranged that they can be set in a perpendicular position for spraying grape vines, and can also be used for spraying currants, berries and other small fruits. It is pushed the same as any wheelbarrow, the revolution of the wheel doing the work. The wheel is 24 inches in diameter, and the tire is 3 inches wide. Weight, 90 lbs. **Price,** complete, including everything shown in the engraving, **$20.00.** This machine can be fitted for attaching a whiffletree for horse power, at an additional cost of **$2.00.**

Norton's Plant Duster.

This is one of the best articles for dusting potatoes, vines, etc., with powder insecticides. A slight jolting movement distributes the dust in a fine cloud. It is made scientifically correct with a cylinder air-chamber projecting above the powder so that it never clogs. Another point of merit is the projecting dust guard, which prevents the powder from escaping beyond the plant being treated. Price, **$1.00** each.

LITTLE GIANT POWDER GUN.

For applying any dry powder, such as Paris green, London purple, Hellebore, Insect powder, Lime, Plaster, etc., on plants or trees, the powder being evenly distributed over a wide space and with the least possible waste; the work being more rapidly done than by any other known implement. It is 24 inches long, with extra tubes for dusting trees, and holds one quart of powder. By turning the crank a fan is rapidly revolved, which forces a current of air through the tubes that carries with it a small portion of powder. The quantity may be increased or diminished, as desired. Price, complete, **$4.00.**

WOODASON'S FLUID AND POWDER BELLOWS
are not the cheapest, but are thoroughly good and made for wear.

WOODASON'S DOUBLE CONE POWDER BELLOWS.
For Dusting Plants with Dry Powders for Insects and Fungus.

This bellows can be held in any direction without wasting powder, as it regulates its own supply, and it does not clog up. The best article ever invented for destroying insects in the conservatory, garden, orchard or field; it will kill every bug upon one acre of potatoes, under the leaves as well as on top, in an hour. (*See cut.*) **Price, $2.75.**

WOODASON'S SINGLE CONE POWDER BELLOWS.
Price, large size, **$1.75.** Small size (for conservatory and house use), **$1.00.**

WOODASON'S FLUID VAPORIZING BELLOWS.
For Spraying Plants with Fluid Decoctions for Insects and Fungus.

This throws a spray as fine as mist, rendering the use of strong solutions perfectly safe on tender foliaged plants; it forces the fluids into every crevice, without using half the quantity required by syringe or sprinkler. (*See cut.*) **Price,** large size, **$2.00.** Small size (for house use), **$1.25.**

WOODASON'S FLUID VAPORIZING BELLOWS.

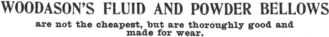

WOODASON'S DOUBLE CONE POWDER BELLOWS.

THE Asbestos Torch.

Attach the torch to the end of a pole of suitable length; saturate with kerosene oil, light and hold under the caterpillars' nests and pass quickly along the branches and around the trunk of the tree where the insects lodge. The heat instantly destroys the insects and will in no way injure the tree. Price, 50c. each; or by mail, 60c.

GREENHOUSE FUMIGATORS, FOR SMOKING PLANTS.

EUREKA FUMIGATORS.

For fumigating greenhouses with dampened tobacco stems; made of galvanized sheet iron; a damper regulates the draft; no danger of fire; no ashes or litter. No. 1, 12 ins. high, $1.50; No. 2, 16 ins. high, $1.75; No. 3, 20 ins. high, $2.00; No. 4, 24 ins. high, $2.75.

PERFECTION FUMIGATORS.

MAKE MOIST SMOKE.

This fumigator will last for years; it has a water tank between the fire and the outside, preventing burning out. The tank should be filled with tobacco water, from which a vapor arises and mixes with the dry smoke from the stems, producing a dampened smoke more dense and less injurious to delicate foliage than from any other fumigator made. Outside fumigating can also be done with the Perfection; the cast iron lid is made so a hose can be attached, and all the outlets for the smoke (except through the hose) can be closed off. No. 1 holds one peck of stems, $3.00; No. 2 holds half bushel of stems, $3.75; No. 3 holds three-quarters bushel stems, $4.50.

Eureka Fumigator.

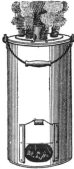

Perfection Fumigator.

═══ HENDERSON'S BRASS SYRINGES. ═══

These Syringes are applicable for all horticultural purposes in the conservatory and garden. They are fitted with Caps or Roses for ejecting water in one stream or dispersing it in a fine or coarse spray, as required. Specially adapted for applying fluid insecticides.

No. A. Length of barrel, 12 in.; diam., 1 in.; 1 spray rose and 1 jet, $2.00.

No. H. Sheet brass Syringe, with fixed spray rose; barrel 16 ins. long; 1½ ins. diameter, $2.25.

No. 10. Barrel 18 ins. long; diam., 1½ ins.; 1 coarse and 1 fine spray rose and 1 stream jet, with patent valves and elbow joint for sprinkling under the foliage, $5.50.

No. G. Barrel 16 ins. long; 1½ ins. diam.; 1 spray rose and 1 stream jet, side attachment and elbow joint for sprinkling under foliage, $4.00.

No. 2. Barrel 14½ ins. long; diam., 1 1/16 in., 1 coarse and 1 fine spray rose and 1 jet, side attachment, $3.50.

No. 11. Same as No. 10, without patent valves, $4.50.

THE "STOTT" PATENT SYRINGE.

X

A large Syringe, the barrel measuring 17 inches long by 4¾ inches in circumference; made of heavy solid brass, highly finished, with one spray rose and one jet, with attachments to fit on the side and patent quick-filling valve and an elbow joint to spray under the foliage. In addition to its value as an ordinary Syringe there is a 3-inch detachable chamber (see Fig. **X**), intended to hold Insecticides, such as Whale Oil Soap, Fir-tree Oil Soap or Tobacco Soap, through which the water is forced, thus automatically mixing and spraying at one operation. The Syringe, of course, can be used either with or without this chamber.

Price, $3.50 each.

"Vermorel" Spray Nozzle.

One of the best for close range. Single spray, 85c.; double spray, $1.50; triple spray, $2.25.

No. 1.   No. 2.   No. 3.

SPRAY NOZZLE ATTACHMENTS.

No. 1. **Pole Connection** so nozzle may be elevated........50c.

No. 2. **Connection** for ¾ in. hose and ¼ in. pipe or pipe threaded nozzle................25c.

No. 3. **Brass "L"** for under-spraying............................25c.

The "Stott" Spray Nozzle.

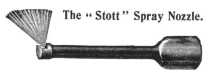

Gives a spray as fine as mist. Well diffused for insecticide purposes; it sprays under the leaves as well as over. Very economical of fluid. It cannot choke or get out of order. Price, single spray for ¾ in. connection (see cut), 50c.; by mail, 60c. Double spray, ¾ in. connection (see cut), 75c.; by mail, 85c.

The "Bordeaux" Nozzle.

SPRAY AND STREAM.

A combination spraying, sprinkling or solid stream. It throws solid stream or fan-shaped spray adjustable to any fineness; is readily degorged by turning the cock handle. It will also throw a long distance coarser spray for spraying very large trees; or it may be shut off altogether. Price for either ¼ in. pipe or ½ or ¾ in. hose, 75c., or by mail, 80c.

The "Numyr"   SPRAY AND STREAM.

A splendid new nozzle, making a fan-shaped spray as fine as mist. Can be regulated to any degree of coarseness, or will throw a straight stream. All changes by simply turning a thumbscrew. Price, for ¼ in. pipe or ¾ in. hose, 50c. each; for 1 in. hose, 60c. (postage, 5c. each extra).

INSECTICIDE APPLIANCES FOR HOUSE PLANTS.

TOBACCO SMOKE PUFFER.

FOR SMOKING WINDOW PLANTS.

Remove the cap, fill the box with any cheap smoking tobacco, packing it tightly. Light as you would a pipe, alternately pressing and relaxing the bulb. Apply the light when air is drawn in and remove it when exhausting the bulb. Use match or wax taper. When well lighted, put on cap. Puff well under the leaves and the green fly will drop off. When through using lay it down and the light will die out. Price, 25c. each, postpaid.

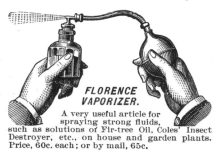

FLORENCE VAPORIZER.

A very useful article for spraying strong fluids, such as solutions of Fir-tree Oil, Coles' Insect Destroyer, etc., on house and garden plants. Price, 60c. each; or by mail, 65c.

JUMBO POWDER GUN.

For applying insect powders on plants in the house or small gardens; it will hold about 4 ozs. of powder, which is ejected and distributed by a pressure with the thumb. The "Jumbo" we consider the best small powder distributer on the market. Price, 20c. each; or by mail, 25c. each.

The Food Values and Fertilizing Ingredients in Garden, Field and Fruit Crops.

* Ash contains lime, magnesia, potash, iron, etc., and forms bone.
⊙ Protein makes lean flesh, blood, skin, muscles, horn, hair, wool, the casein and albumen of milk, etc.
□ Fibre (Cellulose). The framework of plants and the least digestible constituent.
△ Nitrogen Free Extract and ⋎ Fat include starch, sugar, gum, wax, etc., and produce heat, energy, fat, etc.

What a good average crop should produce and consume per acre.

The Fertilizer and quantity required to supply the three essential Plant Foods.

POUNDS OF FOOD INGREDIENTS and FERTILIZING INGREDIENTS CONTAINED IN 1,000 POUNDS OF THE UNDERMENTIONED GARDEN, FIELD AND FRUIT CROPS. (and fractions of lbs.)

A GOOD CROP ON ONE ACRE YIELDING / WOULD CONSUME

THE FERTILIZER RECOMMENDED and the quantity per acre required to supply the probable SOLUBLE deficiency in the soil of the three essentials, viz.: Nitrogen, Phosphoric Acid and Potash.

Water.	Ash. *	Protein. ⊙	Fibre. (Cellulose.) □	Free Ext. △	Fat. ⋎	Nitrogen.	Phosphoric Acid.	Potash.	Crop	Bushels and Tons.	Total Weight.	Nitrogen.	Phosphoric Acid.	Potash.	Use	Lbs. per acre	
lbs.	lbs.	lbs.	lbs.	lbs.	lbs.	lbs.	lbs.	lbs			lbs.	lbs.	lbs.	lbs.			
...	..	..	...	...	...	4.1	1.9	3.9	**Garden Vegetables** { Average of 20 kinds.	.....	24,357	99.8	46.2	94.9	A or L	1,000 to 1,500	Broadcast before harrowing, or 200 lbs. of it may go in drills.
160.	62.	143.	250.	427.	22.	23.	5.3	14.6	**Alfalfa,** hay	5 tons.	10,000	230.	53.	146.	E, I or M	500 to 1,000	Broadcast before harrowing.
799.	4.1	4.9	11.6	120.1	6.5	1.3	.3	1.9	**Apple,** fruit	600 bu.	28,800	36.4	8.6	54.7	A or P	500 to 1,000	Scatter in circle, around each tree.
933.	5.	18.3	7.4	25.5	2.5	3.2	.9	1.2	**Asparagus,** sprouts	.....	12,500	40.	11 2	15.	A, L or S	1,000 to 2,000	Broadcast and rake in.
145.	17.	118.	95.	597.	5.	16.	5.6	2.8	**Barley,** grain	30 bu.	1,440	23.	8.	4.	E, I or M	500 to 1,000	Broadcast before harrowing.
143.	45.9	...	...	...	...	6.4	1.9	10.7	" straw	.....	2,000	12.8	3.8	21.4			
732.	26.	30.	73.	115.	10.	2.9	1.5	5.3	**Beans,** Soja, green fodder	10 tons.	20,000	58.	30.	106.	A or L	600 to 1,200	Broadcast before harrowing.
113.	72.	154.	223.	386.	52.	23.2	6.7	10.8	" " dry "	5 tons.	10,000	232.	67.	108.			
872.	27.6	22.	19.2	75.2	3.7	39.	9.7	12.1	**Beans,** Garden, string	25 bu.	1,500	58.5	14.5	18.1	A or L	600 to 1,200	Broadcast before harrowing, or 200 lbs. of it may go in the drills before sowing.
166.	40.2	...	...	...	...	...	3.9	12.8	" " straw	.....	2,800	...	10.9	35.8			
877.	11.3	15.3	8.8	79.4	1.4	2 4	.9	4.4	**Beets,** roots	400 bu.	24,000	57.6	21.6	105.6	A, L or S	600 to 1,200	
905.	14.6	...	...	...	..	3.	1.	4.5	" leaves	4 tons.	8,000	24.	8.	36.			
889.	5.8	9.4	24.6	50.3	20.8	1.5	.9	2	**Blackberries,** fruit	2,500 qts.	2,500	3.7	2.2	5.	A or P	1,000 to 2,000	Broadcast and cultivate or rake in.
140.	11.8	100.	87.	645.	22.	14.4	5.7	2.7	**Buckwheat,** grain	30 bu.	1,440	20.7	8.2	3.8	E, I or M	300 to 600	Broadcast before harrowing.
160.	51.7	52.	43.	351.	13.	13.	6.1	24.2	" straw	1 ton.	2,000	26.	12.2	48.4			
900.	9.6	23.9	14.7	38.5	3.7	3.	1.1	4.3	**Cabbage**	.....	50,000	150.	55.	215.	A or N	1,000 to 2,000	Broadcast before harrowing, all but 200 lbs. to go in hills, and 200 lbs., scatter around plants later.
904.	8.1	16.2	10.2	49.4	7.9	4.	1 6	3.6	**Cauliflower**	.....	30,000	120.	48.	108.	A or N	1,000 to 2,000	
850.	8.2	11.4	12.7	75.6	4.2	2.2	1.1	3.	**Carrot,** roots	500 bu.	25,000	55.	27.5	75.	A or L	1,000 to 2,000	Broadcast before harrowing, or 200 lbs. of it may go in the drills.
822.	23.9	...	...	...	...	5.1	1.	2.9	" tops	.....	6,000	30.6	6.	17.4			
841.	17.6	..	...	...	...	2.4	2.2	7.6	**Celery**	.....	30,000	72.	66.	228.	A or L	1,000 to 2,000	
825.	3.9	11.	2.4	111.	8.4	...	.6	2.	**Cherries,** fruit	350 bu.	14,000	...	8.4	28.	A or P	500 to 1,000	Scatter in circle, around each tree.
165.	68.4	107.	245.	336.	39.	24.5	6.9	25.3	**Clover,** Red, cured hay	2½ tons.	5,000	122.	34.5	126.	E, I or M	400 to 800	Broadcast before harrowing.
167.	50.7	152.	272.	366.	28.	19.5	3.6	11.7	" Crimson, "	2¼ tons.	4,500	87.7	16.2	52.6			
90.	12.4	103.	22.	704.	5.	14.1	5.7	4.7	**Corn,** Field, ears	125 bu.	8,750	123.7	49.8	41.1	A or O	400 to 800	Broadcast before harrowing, 200 lbs. may be reserved and be put in the hills or drills before planting.
282.	37.4	38.	197.	319.	11.	11.2	3.	13.2	" " stover	4 tons.	8,000	89.6	24.	105.6			
829.	10.4	20.	43.	121.	7.	4.1	1.5	3.8	" Fodder, green	18 tons.	36,000	147.6	54.	136.8			
422.	27.	45.	143.	347.	16.	17.6	5.4	8.9	" " dry	6 tons.	12,000	211.	64.8	106.8			
811.	5.8	28.8	5.4	129.3	9.5	3.4	.6	2.3	" Sweet, green ears	75 bu.	5,250	17.8	3.1	12.			
835.	18.1	19.	67.	155.	9.	4.6	1.1	3.2	" " " stover	5 tons.	10,000	46.	11.	32.			
956.	5.8	8.1	6.9	18.3	2.2	1.6	1.2	2.4	**Cucumber,** fruit	.....	24,000	38.4	28.8	57.6	A or L	1,000 to 2,000	Broadcast, or 200 lbs. of it in hills.
871.	4.1	...	...	...	...	...	.9	1.9	**Currants,** Red, fruit	100 bu.	4,000	...	3.6	7.6	A or P	600 to 1,200	Broadcast and cultivate or rake in.
929.	5.	11.5	7.7	43.4	3.1	...	...	...	**Egg Plant,** fruit	.....	20,000	...	...	...	A or L	1,000 to 2,000	Broadcast, or 200 lbs. of it in hills.
830.	8.8	29.4	37.	568.	5.6	1.7	1.4	5.	**Grape,** fruit	3 tons.	6,000	10.2	8.4	30.	A or P	600 to 1,200	Broadcast and cultivate or rake in.
782.	21.1	30.	100.	175.	120.	7.2	1.9	8.1	**Grass,** Mixed Pasture, green	6 tons.	12,000	86.4	22.8	97.2	E, J or R	400 to 800	Broadcast before harrowing, for seeding or top dressing in spring or fall.
160.	76.	74.	272.	421.	25.	19.1	5.1	22.3	" " " hay	3 tons.	6,000	114.6	30.6	133.8			
910.	12.7	20.1	12.7	42.9	.9	4.8	2.7	4.3	**Kohl Rabi**	500 bu.	25,000	120.	67.5	107.5	A or L	800 to 1,400	
943.	10.3	14.1	7.4	21.8	3.8	2.2	1.	3.9	**Lettuce**	.....	15,000	33.	15.	58.5	A or L	1,000 to 2,000	Broadcast before harrowing, or 200 lbs. of it may go in drills.
873.	12.2	14.	9.	55.	2.	1.9	.9	3.8	**Mangels,** roots	25 tons.	50,000	95.	45.	1?0.	A or S	1,000 to 2,000	
905.	14.6	...	...	...	...	3.	1.	4.5	" leaves	.....	18,000	54.	18.	81.			
877.	9.1	9.3	21.3	184.	1.8	...	...	...	**Melon,** Musk, fruit	5 tons.	10,000	...	...	...	A or L	1,000 to 2,000	Broadcast and plow in one-half; harrow in one-fourth, remainder scatter in hills.
916.	6.2	10.1	5.5	56.4	7.2	...	...	...	**Melon,** Water, fruit	8 tons.	16,000	...	...	...			
870.	12.	26.	110.	187.	10.	6.1	1.9	4.1	**Millet,** Golden, green	8 tons.	16,000	97.6	30.4	65.6	E, I or M	400 to 800	Broadcast before harrowing, or up to 300 lbs. may be drilled with the seed.
98.	45.	...	...	...	...	12.8	4.9	16.9	" " hay	3 tons.	6,000	76.8	29.4	101.4			
750.	19.	24.	70.	131.	6.	5.3	2.	3.4	" Japanese, green	15 tons.	30,000	159.	60.	102.			
104.	58.	51.	301.	433.	18.	11 1	4.	12.2	" " hay	6 tons.	12,000	133.	48.	146.			
143.	26.7	118.	95.	597.	50.	17.6	6.8	4.8	**Oats,** grain	50 bu.	1,600	28.1	10.8	7.6	E, I or M	400 to 800	
143.	61.6	40.	370.	424.	23.	5.6	2.8	16.3	" straw	.....	2,600	14.5	7.2	42.3			
860.	7.4	14.	6.9	95.3	2.6	2.7	1.3	2.5	**Onion,** bulbs	500 bu.	28,000	75.6	36.4	70.	A or L	1,200 to 2,000	Broadcast before harrowing.
877.	6.3	12.	...	...	...	1.9	.8	4.8	**Orange,** fruit	.....	112,000	212.	89.6	537.	A or Q	1,500 to 3,000	Scatter around each tree.
793.	10.	13.5	5.3	160.9	6.6	5.4	1.9	6.2	**Parsnip,** roots	650 bu.	30,000	162.	57.	186.	A or L	1,000 to 2,000	Broadcast and harrow in, or 200 lbs. of it may go in the drills.
831.	25.9	...	...	...	...	2.9	.8	2.5	" tops	.....	10,000	29.	8.	25.			
884.	3.4	...	...	...	...	...	.5	2.5	**Peaches,** fruit	900 bu.	30,000	...	15.	75.	A or P	400 to 800	Scatter in circles, 15 feet diameter, around each tree.
582.	19.3	...	...	...	...	9.	2.2	5.	" new tree growth	.....	6,000	54.	13.2	30.			
143.	23.4	270.4	39.	575.5	15.8	35.8	8.4	10.1	**Peas,** Garden, seeds	25 bu.	1.500	53.7	12.6	15.1	A or L	1,000 to 2,000	Broadcast before harrowing, or 200 lbs. of it may be applied in the drills.
160.	43.1	...	...	...	...	10.4	3.5	9.9	" " straw	.....	3,600	37.4	12.6	35.6			
788.	17.	24.	48.	71.	4.	2.9	1.	3.1	**Peas,** Field, cow, green plants	8 tons.	16,000	46.4	16.	49.6	E or I	400 to 800	
107.	75.	166.	201.	422.	29.	19.5	5.2	14.7	" " " dry plants	4 tons.	8,000	156.	41.6	117.6			
831.	5.4	5.6	27.3	114.	7.9	.6	.5	1.8	**Pears,** fruit	600 bu.	28,800	17.2	14.4	51.8	A or P	800 to 1,600	Scatter in circle, 15 feet diameter, around each tree.
838.	4.9	11.3	...	...	...	1.8	.4	2.4	**Plums,** fruit	500 bu.	20,000	36.	8.	48.	A or P	800 to 1,600	
750.	9.5	21.	6.	173.	1.	3.4	1.6	5.8	**Potatoes,** tubers	300 bu.	18,000	61.2	28.8	104.4	S or T	800 to 1,600	Broadcast before harrowing, or 500 lbs. of it may be scattered in the rows before planting, and 500 lbs. between rows at first hoeing.
770.	19.7	...	...	...	...	4.9	1.6	4.3	" tops	.....	6,000	29.4	9.6	25.8			
758.	7.4	14.2	12.3	247.	3.5	2.4	.8	3.7	**Potatoes, Sweet,** tubers	350 bu.	19,250	46.2	15.4	71.2	A or L	1,000 to 2,000	
415.	57.9	76.6	136.	292.	21.1	4.2	.7	7.3	" " vines	.....	10,000	42.	7.	73.			
900.	4.4	11.1	14.9	43.4	1.6	1.1	.7	.9	**Pumpkins,** fruit	8 tons.	16,000	17.6	11.2	14.4	A or L	800 to 1,600	Broadcast, or 200 lbs. of it in hills.
818.	9.5	9.9	28.8	...	...	1.5	4.8	3 5	**Raspberries,** fruit	2,000 qts.	2,000	3.	9.6	7.	A or P	1,000 to 2,000	Broadcast and cultivate or rake in.
933.	4.9	...	...	...	...	1.9	4.5	1.6	**Radish**	3 tons.	6,000	11.4	27.	9.6	A or L	800 to 1,600	Broadcast before harrowing.
891.	10.6	11.8	12.5	76.6	1.5	1.9	1.2	4.9	**Ruta Baga,** roots	800 bu.	40,000	76.	48.	196.	E, I or M	400 to 800	Broadcast before harrowing, or 200 lbs. of it may go in the drills.
898.	11.9	...	...	...	...	3.	.9	2.8	" leaves and tops	4 tons.	8,000	24.	7.2	22.4			
143.	17.9	106.	17.	725.	17.	17.6	8.5	5.8	**Rye,** grain	32 bu.	1,792	31.5	15.2	10.3	E, I or M	400 to 800	Broadcast before harrowing, or half of it may be drilled in.
143.	38.2	30.	389.	466.	12.	4.	2.5	8.6	" straw	.....	3,800	15.2	9.5	32.6			
801.	13.7	13.	61.	116.	5.	3.3	.8	3.6	**Sorghum Plant,** green	12 tons.	24,000	79.2	19.2	86.4	S	1,000 to 2,000	Broadcast before harrowing, or 500 lbs. of it in furrows.
...	...	...	...	...	...	...	...	...	" " dry	4 tons.	8,000	...	...	...			
922.	19.6	21.	6.7	23.8	4.9	4.9	1.6	2.7	**Spinach**	200 bbls.	12,000	58.8	19.2	32.4	A or L	1,000 to 2,000	Broadcast before harrowing.
948.	4.1	9.2	10.4	80.5	1.8	...	...	...	**Squash,** fruit	8 tons.	16,000	...	...	...	A or L	800 to 1,600	Broadcast, or 200 lbs. of it in hills.
869.	10.4	18.	9.	98.	1.	2.2	1.	4.8	**Sugar Beet,** roots	18 tons.	36,000	79.2	36.	172.8	A, S or T	800 to 1,600	Broadcast before harrowing, or 500 lbs. of it in the drills.
868.	12.5	...	...	...	...	3.5	1.	3.4	" " leaves	.....	10,000	35.	10.	34.			
908.	5.2	9.5	14.3	55.	6.8	1.5	1.	2.6	**Strawberry,** fruit	5,000 qts.	5,000	7.5	5.	13.	A or P	1,000 to 2,000	Broadcast and cultivate or rake in.
180.	140.7	...	...	...	...	34.8	6.6	40.7	**Tobacco,** Dry, leaves	.....	1,260	43.8	8.3	51.2	U	1,000 to 2,000	Broadcast one-half before plowing the rest before harrowing
180.	64.7	...	...	...	...	24.6	9.2	28.2	" " stalks	.....	1,100	27.	10.1	31.			
895.	10.1	11.4	11.5	62.7	1.8	1.8	1.	3.9	**Turnip,** roots	600 bu.	33,000	59.4	33.	128.7	A or L	800 to 1,600	Broadcast before harrowing, or 500 lbs. of it may go in the drills.
898.	11.9	...	...	...	...	3.	.9	2.8	" leaves	.....	9,000	27.	8.1	25.2			
940.	5.7	9.1	7.5	38.	4.3	1.7	.4	3.6	**Tomato,** fruit	400 bu.	20,000	34.	8.	72.	A or L	800 to 1,600	Broadcast before harrowing, or 300 lbs. of it may go in the drills.
836.	30.	...	...	...	...	3.2	.7	5.	" plants	.....	15,000	48.	10.5	75.			
144.	16.8	119.	18.	719.	21.	20.8	7.9	5.2	**Wheat,** grain	30 bu.	1,800	37.4	13.9	9.3	E, I or M	400 to 800	Broadcast before harrowing or drill in.
143.	46.	34.	381.	434.	13.	4.8	2.2	6.3	" straw	.....	3,600	17.2	7.9	22.6			

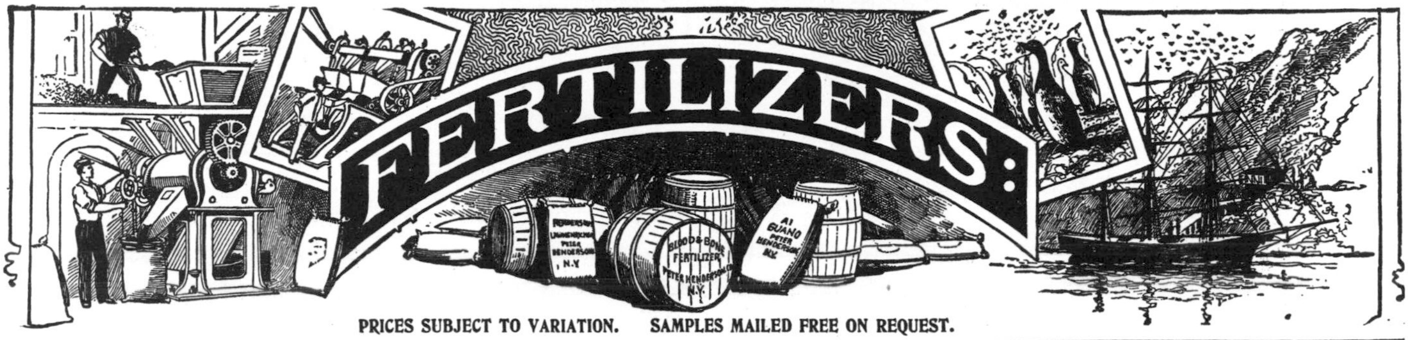

PRICES SUBJECT TO VARIATION. SAMPLES MAILED FREE ON REQUEST.

Referred to on opposite page as LETTER		PRICES. Free on board cars, N. Y. Bag of 100 lbs.	Bag of 200 lbs.	Ton, 2,000 lbs.	Nitrogen in 1,000 lbs.	Phosphoric Acid. Available in 1,000 lbs.	Potash. Available in 1,000 lbs.
	THE HENDERSON LAWN ENRICHER. (*See description and prices on page 59.*)						
	HENDERSON'S SUPERIOR FERTILIZER, FOR HOUSE PLANTS. Price, 1 lb. package, sufficient for 25 ordinary sized plants for 1 year, 20c., or by mail, 35c. A safe, clean fertilizer, free from disagreeable odor, prepared especially for feeding plants grown in pots. It is a wonderful invigorator, producing luxuriant, healthy growth, and larger and more brilliant flowers. Detailed directions on each package.						
A	**HENDERSON'S GARDEN FERTILIZER, FOR VEGETABLES AND FLOWERS.** Prices, 5 lb. pkg., 25c.; 10 lb., 45c.; 25 lb., $1.00; 50 lb., $1.75. Contains all the elements in well-balanced proportions needed by vegetables and flowers for their quick growth, early maturing, and perfect development; immediately commences to feed the plants, stimulating them into healthy, luxuriant growth, and it continues to feed them until the end of the season. Highly concentrated, fine, dry and free from objectionable odor, easily applied, either before or after planting. If especially fine vegetables or flowers are desired, two or three additional applications will produce magnificent results. *A 10-lb. package is sufficient for a space 15 x 20 ft. for one application, or 500 lbs. to 600 lbs. for an acre.*	$3.00	$5.00	$45.00	49 lb.	90 lb.	25 lb.
B	**GROUND BONE,** Ground raw, Pure.......... Unsifted, containing fine and coarser particles just as ground; the finer feed the plants immediately, while the coarser keep up the supply for a long period. For top-dressing grass, ¼ to ½ ton acre; garden and field crops, ¾ to 1 ton.	2.25	4.00	35.00	33 lb.	80 lb.	1 lb.
C	**BONE MEAL,** Ground raw, Pure.......... Prices, 1 lb. pkg., 10c.; 5 lb., 30c.; 10 lb., 50c.; 25 lb., $1.00; 50 lb., $1.50. Ground very fine, and, therefore, more quickly beneficial to plants than the coarser grades. This is also used as "**feeding meal**" for poultry and animals. For top-dressing grass, ¼ to ½ ton acre; garden and field crops, ¾ to 1 ton; rose beds, pot plants, etc., 1 part to 50 soil.	2.50	4.25	37.00	37 lb.	85 lb.	1 lb.
D	**CRUSHED BONE,** Raw, Pure.......... One-eighth, quarter and half-inch pieces, very slow in action, but permanent; indispensable in preparing soil for planting grape vines, small fruits, trees, etc. For foundation of permanent grass land and lawns, ½ to 1 ton acre; trees, grapes, etc., 2 to 4 qts. each.	2.50	4.50	42.00	23 lb.	80 lb.	1 lb.
E	**BONE SUPERPHOSPHATE..........** Benefits crops quickly, promotes early maturity, affords constant nourishment during the entire season. In case of rotation of crop, it is nearly as valuable for the last as the first. Quantity required: For garden and field crops, ½ to 1 ton per acre; top-dressing grass, ¼ to ½ ton acre.	2.00	3.50	32.00	21 lb.	100 lb.	25 lb.
F	**BLOOD AND BONE..........** Superior for immediate as well as prolonged results; a splendid source of nitrogen and phosphoric acid. Quantity required: Garden and field crops, ½ to 1 ton per acre; broadcast or in drill, 300 to 500 lbs.	2.25	4.00	36.00	49 lb.	110 lb.	....
G	**SHEEP MANURE,** Pulverized.......... A pure, natural manure and most nutritious for all kinds of plants; its effect is immediate; it makes a safe and rich liquid manure. Quantity required, ¾ to 2 tons per acre.	2.00	3.50	30.00	21 lb.	15 lb.	25 lb.
H	**COTTON SEED MEAL..........** The elements of plant food are quickly available, and feed and push the crop along vigorously; it, however, is not a lasting fertilizer, and better adapted for early crops. Quantity required, ½ to 1 ton per acre broadcast; 500 to 1,000 lbs. in drill.	2.25	4.00	35.00	49 lb.	20 lb.	15 lb.
I	**MAPES' COMPLETE MANURE, "A" BRAND..........** Specially adapted for hill or drill use on all crops, particularly in connection with stable manures. Fine for peas, beans, buckwheat and turnips.	2.00	3.75	35.00	25 lb.	110 lb.	25 lb.
J	for **GENERAL USE..........** A substitute for stable manure; for use on all soils for all crops with or without stable manure; special for oats, millets, vegetables, melons, tomatoes and seeding with or without grain.	2.25	4.00	38.00	33 lb.	90 lb.	40 lb.
K	for **HEAVY SOILS..........** For any soils where small quantity of potash but large quantities of ammonia and phosphoric acid are required. Safe to use around young plants, nursery stock, strawberry vines; very forcing; special for early turnips. Use ¼ to ½ ton per acre.	2.50	4.25	39.00	10 lb.	80 lb.	[illegible]
L	for **VEGETABLES** or Light Soils.......... For truck, early and late vegetables, onions, celery, tomatoes, etc., on all soils, and also for oats, hops and barley on light soils. Use for vegetables, etc., ½ to 1 ton per acre; grass, grain, etc., 400 to 800 lbs.	2.50	4.25	41.00	49 lb.	60 lb.	60 lb.
M	for **FARM CROPS,** Cereal Brand.......... For wheat, rye, corn, oats, buckwheat and all farm crops, particularly where farm manures are used.	1.75	3.00	28.00	16 lb.	70 lb.	30 lb.
N	for **CABBAGE and Cauliflower..........** Contains the elements of plant food in well-balanced proportions to produce maximum crops; of superior quality and whiteness and firmness. Use 800 to 1,600 lbs. per acre.	2.25	4.00	38.00	41 lb.	60 lb.	60 lb.
O	**CORN MANURE..........** For field and sweet corn, fodder corn, millets, late cabbage, late turnips and seeding to grass. On good land in drills, 200 to 400 lbs.; on poorer light lands broadcast, 600 to 800 lbs. per acre.	2.25	3.75	36.00	25 lb.	90 lb.	60 lb.
P	**FRUIT AND VINE MANURE..........** For insuring fruiting power and quality of fruit for grapes, pears, apples, plums, strawberries and all small fruits; effects slow, but lasting. Use 600 to 1,000 lbs. per acre.	2.25	4.00	37.00	16 lb.	70 lb.	100 lb.
Q	**FRUIT-TREE AND ORANGE MANURE..........** Promotes wood growth and fruiting; for all fruit-trees. Use 600 to 2,000 lbs. per acre.	2.25	4.00	37.00	33 lb.	80 lb.	30 lb.
R	**GRASS AND GRAIN,** Spring Top-dressing.......... For all kinds of grasses and pastures, mowing lands, timothy clover, wheat, oats, rye and all grain crops. On good land, 200 to 600 lbs., and poor land, 600 lbs. per acre.	2.50	4.25	40.00	49 lb.	50 lb.	70 lb.
S	**POTATO MANURE..........** For Irish and sweet potatoes, also asparagus, tomatoes, sugar beets, fruits, sorghum, sugar cane and sugar corn. 600 to 1,000 lbs. per acre.	2.50	4.25	39.00	37 lb.	80 lb.	60 lb.
T	**POTATO MANURE,** Economical Brand.......... Not as forcing as No. L, yet adapted for all vegetables where large crop and superior quality, particularly in sugar and starch, is desired.	2.25	3.85	35.00	33 lb.	45 lb.	80 lb.
U	**TOBACCO MANURE,** Wrapper Brand.......... For growing superior quality of leaf, particularly for wrappers. Use ½ to 1 ton per acre.	2.65	4.35	39.00	62 lb.	40 lb.	105 lb.
V	**ASHES,** Canada Hardwood, Unleached.......... Very beneficial for garden, field and fruit crops requiring potash; excellent top-dressing for grass; improves the texture of the soil and drives away insects. Quantity required, 1 to 2 tons per acre.	1.25	2.00	18.00	....	10 lb.	60 lb.
V V	**SULPHATE OF POTASH,** High Grade.......... The highest grade and purest form of agricultural potash used in the preparation of fertilizers.	3.50	6.50	Apply.	....	....	500 lb.
W	**KAINIT** (German Potash Salt).......... In addition to its potash value, it contains about 30 per cent salt; the combination of these two ingredients renders it very useful in destroying insects in the soil, such as slugs, cutworms, wireworms, maggots, etc.	2.00	3.50	Apply.	....	....	130 lb.
X	**NITRATE OF SODA..........** Prices, 5 lb. pkg., 30c.; 10 lb., 50c.; 25 lb., $1.25; 50 lb., $2.00. Valuable solely for its nitrogen; it is very stimulating, quick in action, and hastens earliness and maturity; must be used in connection with other fertilizers. Quantity required: Being extremely soluble it should not be applied until the plants are above ground, when 100 to 500 lbs. per acre, mixed with wood ashes or land plaster, for convenience in applying, are generally used.	3.50	6.00	55.00	150 lb.	....	....
Y	**BONE BLACK, DISSOLVED..........** Esteemed for applying available phosphoric acid in the preparation of fertilizers.	1.75	3.00	Apply.	....	160 lb.	....
Z	**LAND PLASTER..........** Per barrel of 250 lbs., net weight, $1.50.	....	....	11.00	....	....	....